RSA Cracked

German Navarro

Copyright © 2020 German Navarro

All rights reserved.

ISBN:9781074921507

DEDICATION

To them who persevere, no matter what.

CONTENTS

	Acknowledgments	i
	Preface	1
1	Zero	6
2	Scales of measurement	18
3	Generating the prime numbers	31
4	Large primes	43
5	The twins	56
6	Cryptography	66
7	Reflections	91

ACKNOWLEDGMENTS

I acknowledge Nature for its stiff refusal to get determined by human mind.

Preface.

This book deals with numbers, particularly with the number zero and the prime numbers. The number zero has no doubt played an important role in the development of modern science. Science however has a double meaning in this book: engineering on the one hand, and the search for fundamental variables of Nature on the other. The former is considered to be applied science; the latter is viewed as man's effort to explain the great mysteries of the universe we live in. The author's position is that the number zero "behaves" differently in engineering and in fundamental research.

While dealing with the scales of measurement, special attention is given to the number zero in view of the absolute importance this number has in measurements at ratio level. What would happen to the fundamental research of Nature should we ever get dispossessed of the number zero? In the pages that follow, the author tries to give an answer to this fundamental question.

The reader will also come across a new approach to

prime numbers. After finding a new environment where to look for them, this new approach leads to a new solution, a solution that disregards completely the prime numbers' indivisibility. Individual primes and twins, dispossessed of their quantitative identity become mere rankings in an anonymous ordinal sequence where they are extremely easy to localise. The prime numbers' hideout gets disclosed.

In addition, primes can be generated not only starting from zero but from any place alongside the numerical sequence. No knowledge is needed of the preceding primes. So, one could start generating primes that are greater than say 10^{12} without any knowledge of the primes that are less than 10^{12}.

Besides, the algorithm presented in these pages opens the door to the creation of a data base aimed at storing huge primes. No matter how big a prime number is, even if it consists of thousands of digits, it will take not more than one single bit to get stored in a storage device.

Using the previously mentioned anonymous sequence, a method is developed to swiftly factor the semiprimes used by the mathematicians to seal off the Internet. If the author is right, anyone who has at his disposal the same resources, human and otherwise, that the mathematicians have, could render RSA and similar brands of public key cryptography utterly obsolete.

Before jumping into the subject-matter, a word of caution should be given. The 3 chapters that deal with the prime numbers are aimed in the first place at showing that the primes are not **cardinal numbers** but mere **rankings** in an ordinal array.

As it is very well known, rankings cannot be added up or subtracted from one another, let alone be multiplied or divided by one another. This is why one cannot apply on the primes that are disclosed using the present method the definition used by the mathematicians of all ages. According to them, primes are those **numbers** that are **divisible** only by themselves and by 1. The primes that were found using the present method are **rankings** and as such they are not divisible at all.

In the second place the methodology presented in this book is aimed at disclosing a way to circumvent the classical solution used by the mathematicians of all ages to try to solve the factoring of rather large semiprimes.

The author's conviction that led him to look for the primes in the reign of ordinal "numbers" was intrinsically tied to his conviction that not the prime factors but rather the semiprime itself should be the entity to be sought after. For also a semiprime, if it gets transferred to the ordinal world, ceases to be a cardinal number and becomes there a mere ranking.

Well, being it so that the mathematicians publish the semiprime (and not its prime factors) on the Internet, the reader is led first to grab it from that public dominion, an operation that takes one trillionth of a second to execute. The reader knows that the semiprime concerned has its own, exclusive "habitat" in the world of ordinal things. He is instructed then to go there and take a look, an operation that takes another trillionth of a second to be executed. In the third place the reader is instructed to grab anything he finds there and this is why factoring a semiprime takes no more than 3 trillionths of a second: the semiprime concerned was previously chased away from its natural "habitat" and replaced by one of its prime factors. And that is what the reader finds there.

In order to achieve such a swift solution, one has to consult a data base. And data bases designed to give home to primes tend to be unfeasible due to the sheer amount of numbers they should contain. Besides, reading carefully the last paragraph leads to the conclusion that the data base proposed should contain myriads of semiprimes.

Well, finding semiprimes that are as big as 10^7, 10^8 shouldn't be that difficult. The author is not aware of existing methods designed to isolate all semiprimes from other composites, and then semiprimes of the magnitude of 10^{50}, 10^{60}. To the reader of this book a method is offered to isolate them, to find them, all of them, and to construct with them a data base. Any semiprime that the mathematicians can find multiplying any prime by any other prime, will be there. And the reason why such a data base is feasible is astonishingly simple: the amount of semiprimes present in the numerical system **decreases** dramatically, or exponentially one should say, as the set of so-called natural numbers expands towards infinity. Read it all in Chapter 6.

The author thinks a data base to give home to semiprimes of that magnitude is feasible. The resources though one needs to execute such a daring enterprise should meet one's ambitions. Read about this in the second half of Chapter 6, the place where the author gets deep into the matter. This is the place where he tries to deliver what the title of the present book promises: to crack public key cryptography.

A last word of caution follows for the experts in computer technology. The author does not know how to program a computer, so if you are looking for a short cut to generate myriads and myriads of prime numbers in one trillionth of a second and become the winner of a race that apparently is going on, you will be highly deceived. The author limits himself in this book to suggest a rough sketch of how to arrange things on the computer in order to obtain the desired results rather than to write a guide on how to obtain primes, twins, semiprimes and what not in the most efficient of ways. Computer efficiency should be handled by those who know how to deal with computers, not by those who, like the author, have never written successfully two consecutive lines of code.

1 ZERO

In *The Lost Civilizations of the Stone Age*, Richard Rudgley describes masterfully the way human kind learned how to count. The Middle Eastern tribes whose humble way of life he describes were not in search of the great mysteries of Nature, but were rather tentatively trying to ascertain each individual's share of the commonly possessed livestock.

Those tribes had changed their way of life profoundly. They were no longer nomads, wandering from one water source to the other, hiding from predators, searching for new fertile land to pasture their commonly possessed stock. Instead, they had chosen to settle down in the vicinity of water, both from heaven and earth, water that they desperately needed were they to survive. In the new settlements, the individual human began to count, networks were established, family lines were neatly drawn. The new communities gave birth to "the individual right of property", something unknown while humans were still nomads. They began to concern themselves with mastering the way to express each individual's or family's wealth in terms of the amount of objects or livestock they possessed, and numbers did just that.

Before they learned how to count properly, it is probable that our ancestors had a rudimentary counting system based on intuition. We, modern humans, possess a nature-given, visual skill to count. Look at the sky and observe the birds flying: if it is one, you know it is only one. If it is two, you know it. If it is three, you don't have to count them. If it is four, you'll have to make sure it is four. By about five or six, the problems begin. You have to count them; otherwise you do not know how many birds you saw. Apparently, our ability to count intuitively has a threshold, and quite a low one for that matter.

So, knowing for certain that the human beings studied by Rudgley were absolutely comparable to modern man, there is no reason to believe that their ability to count intuitively differed from ours. It is imaginable therefore that they had distinct words to indicate the amounts of objects that they could discern quantitatively from one another in an intuitive way. If we agree to put this limit at a conservative 6, then it is possible that they had the ability to count from 1 to 6 but that anything equal to or larger than 7 could only have been spoken of as "much" or "several" or "many".

However it might have been, once the former nomadic tribes became sedentary, culture began to evolve and with culture knowledge and with knowledge economics and with economics numbers. It is superfluous to repeat a story so magnificently told by Rudgley. Instead, a schematic view of what happened to numbers will be presented here: the new civilizations did indeed develop a sophisticated counting system, so sophisticated that it only marginally differs from ours, thousands of years later. Numbers were born.

The individual Stone Age man possessed sheep, goats, donkeys, all sorts of other domesticated animals and plants in quantities diverse from those of his neighbours, and so arose the absolute necessity to distinguish 17 goats both from 16 and from 18, as intuitive, eye-counting did not suffice any more to express and differentiate their possessions.

The counting system developed by the Middle Eastern tribes was meant to express *discrete* quantities. The lowest number they knew was therefore 1; zero did not belong to their counting system as zero is the absence of any quantity and the counting system was aimed at expressing quantities, not the absence thereof. Shortly, counting was nothing other than applying numbers directly to the items that were counted almost in a predicative way. "The number of sheep I possess is (equal to) nine", "the number of children my brother has is (equal to) three". The items that were counted were discrete, not divisible: one quarter of a goat or two thirds of an arrow were still waiting to be invented.

We, modern humans have a tendency to think easily about numbers. By the time we reach the age of ten or twelve years, most of us are capable of counting indefinitely, not realising that it could have taken the human race one thousand years to learn how to count from one to one thousand, so to speak. Apparently, our brain was made to deal with numbers, certainly not to invent them.

Another thing we don't realise is the huge difference that exists between *counting* and *measuring*. Although strictly related with one another, they are two different activities of the human mind. Whereas counting, as it has just been stated, applies numbers directly to bare objects, measuring is based on an intermediate social agreement. Measuring presupposes that a society, large or small, has reached agreement on a standard that is to be taken as the unit of measurement. Again, taking the example of the ten or twelve year old child: show him two coins, one of 1 Euro and another of 2. Ask him how much is that and he will give you the right answer with no hesitation: three Euros. What the child does not realise is that he has done two separate mental operations: first of all, he has *counted* the coins, being two, and in the second place he has *measured* each coin's value being one and two Euros, yielding three Euros as the right answer. What has happened in this instance is that the child is aware of the social agreement that has assigned a particular value to a particular piece of metal coined in a particular way, accepts that agreement, evaluates (= measures) each coin's value and counts them.

Counting coins and counting values: they both go: one, two, three ... They seem to be identical, but in reality they belong to two separate worlds. If it took the human race one thousand years to learn how to count from one to one thousand, as earlier suggested, another thousand years were needed to socially accept a vessel of some sort as being the standard of measurement for liquids and grains (again, so to speak).

It is probable that the first standards upon which social agreement was ever reached expressed a *small amount* of "something". That "something" was probably a commodity that had either liquid or granular properties and that therefore had to be contained in a bag or vessel of some sort in order for it to be transported or indeed bartered or even sold.

That such was indeed the case is commonly agreed upon in history. Mesopotamians, Greek, Romans ... they all knew all sorts of standards accepted by society and ratified by law.

What happens mathematically when a society goes over from counting to measuring is that counting no longer is applied to *concrete objects*, indivisible entities that are essentially identical to one another and interchangeable, but rather to some *abstract measurement unit* that represents or contains a standard amount, quantity, weight, length ... that socially has been agreed upon. Society goes over from one, two, three *houses* to one, two, three *feet*. Counting and measuring, they both operate on numbers but yield different answers: after counting, the resulting number gives a particular amount of individual objects whereas after measuring, the result expresses the number of measurement units (or the fraction thereof) that such an amount contains or represents. The *counting system*, earlier applied directly on bare objects, becomes a *measurement system* in that it is a *measure* that has acquired an intermediate role between matter and numbers. Measuring is counting abstract objects, so to speak.

We don't know how exactly fractions did come into existence but it is very well imaginable that fractioning was a consequence of measuring. As it was said, it is probable that the first measurement standards were intended to express or contain small quantities of the substance or commodity being measured. As human settlements grew ever larger and larger, there probably arose the necessity of constructing greater "containers" that were nothing other than a multiple of the smaller ones, the first units of measurement upon which social agreement had been reached. From there to a reverse situation, where the larger unit of measurement becomes the standard and the smaller, original one becomes a "fraction" thereof, a very small step is needed.

During thousands of years numbers were used only for *counting* purposes, but by the time the Mesopotamian culture arose, we know that numbers were also used as an instrument of *measurement*. The step from counting to measuring is a huge one: for the former all one needs is a discrete amount of units to be counted, whereas for the latter social accord is needed: every Roman knew (by approximation) what a *pes* or a *congius* was. The social agreement that is needed to use numbers as an instrument of measurement creates the necessity of the use of fractions: Romans, for instance, knew both small and large units of measurement. Using the smallest units as a departure point, they expressed the larger ones as multiples thereof, whereas departing from the larger units they expressed the smaller ones as a fraction of them.

Speaking in today's terms, while human beings were able only to count they were able to express or manipulate *discrete variables,* whereas with measurement, *continuous variables* appeared, continuity being nothing other than the ability of a variable to acquire not only whole, but also fractional values. Actually, what *measuring* did to *counting* was that the former created a whole range of new (fractional) numbers between any two consecutive digits used in the old counting system but most important of all: measuring opened the infinitely deep gap below 1, the lowest of all numbers that the counting system knew. Modern mathematicians would undoubtedly say that the gap between 0 and 1 was finally covered, an assertion that is only marginally true.

Going back to counting, it is hardly imaginable that its inventors realised how powerful this instrument would become in history. In due course the numbers themselves were to become the object of research. And even as far as these pages are concerned, they will deal with some aspects of our ancestors' counting system that have survived time as well as with some relatively new, questionable additions.

Counting up, to the right side of the number 1 as it were, mathematicians soon concluded not only that they were unable to find an upper limit to counting but that indeed the counting system as such had no upper limit at all. The greatest possible number could never be found: should anyone claim to have found it, all one had to do was to add 1 to it, proving the claim to be false.

Counting down was quite another matter and it all has to do with the number zero. Modern mathematicians are inclined to take zero as the departure point of the counting system, not only to its "right" but also to its "left". Numbers on its right are positive, numbers to its left are negative. Should one try to find the least possible number, one can apply the same reasoning used for positive numbers: you can count down as long as you want to the left of zero, a limit will never be reached.

The problem with zero in modern mathematics was its unquestioned entrance into the kingdom of numbers. What is more, zero was placed at the very centre of the numerical creation surrounded by an infinity of numbers on its right side as well as on its left side. Zero was actually granted the role of "king of numbers", a title that zero certainly does not deserve.

The most fundamental question about zero is: is zero really a number? Obviously, in a *discrete* counting system there is no place for it. Saying, as mathematicians tend to do, that a person that has no sheep, actually possesses zero sheep is an aberration to say the least. As it was earlier stated, the lowest possible number in a discrete counting system is 1. Such a system knows no fractions on the one hand, and on the other, a discrete counting system is meant to count discrete quantities of entities, entities that are indivisible. This all means that, concretely speaking, there is no such thing as a fraction of a sheep or indeed a nonexistent or "empty" sheep to which the "number" zero can be attached to in order to start counting from there. The zero sheep exists only in the mathematician's mind.

Parallel to the question of the existence of zero as a member of the "*counting*" family of numbers runs the question of its existence as a member of the "*measuring*" family. It was earlier stated that measuring introduced fractioning and with fractioning an attempt was made to bridge the infinite gap between one and zero. As it has just been said, mathematicians accepted zero as the "natural" number placed half way between positive and negative amounts. If zero has been rejected as a member of the counting family, then there arises the question of the legitimacy of its existence within the measuring family of numbers. Well, if it is unacceptable to speak of zero sheep, is it legitimate to speak of zero feet, zero pounds, or zero Euros?

At the very face of it, there is nothing wrong with giving a positive answer to these questions. Actually, the tremendous advance of science wouldn't have been possible without the presence of zero in our measurement system. For engineering purposes we can very well pretend to be able to depart from zero speed, zero weight, zero time. And as long as we keep measuring our small world, building instruments, observing the behaviour of matter on Earth and its surroundings, there is nothing wrong with it. The problem arises when "science" jumps from engineering to rampant speculation: big bangs, iterative universes, speed limiting ... all these thoughts can't be conceived without zeroing one or more of the fundamental variables that constitute Nature, be it time, be it energy, be it mass ...

But is it valid to speak of zero time, zero energy, zero light, or zero mass? Mathematicians have absolutely no problem with these questions; neither do they have a problem with their answers.

Yes, they say, it is utterly legitimate to speak of zero time, zero energy, zero mass and so on. But then, what do mathematicians mean when they zero these fundamental variables of Nature? Obviously, while attributing zero to energy, they refer to a situation in which the "amount" of energy indicated is nonexistent, which is inherent to the very definition of zero. There is nothing "there". The only thing that is "there" is emptiness. And, when they speak of zero time there is nothing "there". The only thing that is "there" is also emptiness.

This concept of "emptiness" is inherent to the very essence of measuring. While measuring some particular variable, the human mind concentrates solely on the variable being measured, pretending as it were that the rest of reality does not exist. When we speak of 20 kilograms of mass for instance we implicitly abstract from measuring any other variable and indeed, we even abstract from the presence of any other variable. So, while measuring a particular amount of mass and concluding that this amount is equal to zero we actually conclude that "emptiness" is there. There is nothing "there". Not only no mass, but also no light for we did not intend to measure any light; no time, for we did not intend to measure any time; no energy, for we did not intend to measure any energy ... so, measuring energy and finding a zero amount of it yields the same result as measuring mass and finding a zero amount of it or measuring light and finding a zero amount of it. Generally speaking a zero quantity of any variable that can acquire that value is identical to a zero value of any other variable: zero time is perfectly identical to zero energy, to zero mass, to zero distance.

So, zeroing any fundamental variable of Nature leads to the same result. And this result is emptiness, as zero is emptiness, the negation of everything. Departing from energy for example, zero energy means also zero mass, zero time, zero light, zero gravity ...Zero cannot be "applied" to any variable in Nature without "applying" it to all other variables that constitute the same Nature. This is due to the overwhelming, "destructive" power of zero. Once Nature is "there", even with the tiniest possible "amount" of anything, then and only then has zero been left behind, emptiness is no longer "there". That tiny "amount" of energy, or time, or mass ... lends itself then to be measured, reality is back, and Nature is again present.

After these considerations it becomes clear that choosing zero as a departing point for any abstract, non-engineering related measurement means:

1. Choosing as departure point a "place" outside the universe as it is known to us: a "place" with zero energy, zero light, zero mass, zero gravity ... How *on earth* mathematicians are then able to "jump" from there to this universe, at least to this earth of ours, full of energy, full of light ... remains as yet an unexplained miracle.

2. Since zero energy is the absolute absence not only of energy but also of anything else, it is obvious that zero energy is absolutely equal to zero gravity, to zero light, and so forth, for those "zero states" also represent the absolute absence of everything. Well, taking a step, as mathematicians theoretically do, from zero energy to the smallest possible amount of energy is tantamount to "jumping" from "zero light" to the smallest possible amount of energy, from "zero gravity" to the smallest possible amount of energy, from "zero space" to the smallest possible amount of mass ... in a few words, a "jump" from zero to Nature results in a continuity between energy and light, between energy and time, between space and mass ... obviously, an aberration, to say the least.

Finally, there is a fundamental question regarding the presence of zero within the *measurement* system. As it was earlier stated, measurement made fractioning possible and fractioning did open the way to explore quantities below the number 1, the absolute lowest limit admitted by the *counting* system.

It is well known, and admitted by everybody, that our counting *and* our measuring systems do not know an upper limit, meaning that either the human mind is incapable of finding the greatest possible number or that such a number does not exist. In either case it is far beyond our mental capabilities to get a hold of it.

The question then arises: what about the smallest possible number? Well, once zero got introduced (or rather: smuggled) into our numerical systems, mathematicians simply "flew" over zero and declared the problem of the smallest possible number to be simply the mirror image of the greatest possible one. In other words, the smallest possible number is equal to the greatest possible one with a negative sign attached to it.

In reality, once fractions come into existence the smallest possible number has to be sought below 1, as near to zero as possible. Once fractions have entered the human mind, one cannot disregard the presence of zero as being the absolute absence of any quantity at all. Well, no matter how deep you dig underneath 1, you will never ever be able to lay your hand on the smallest possible number for you will never ever get to zero. That is where the smallest possible number "resides" and not at the very end of the negative numbers as mathematicians state.

From what has just been said one is bound to conclude that zero is as disqualified to enter the family of *measuring* numbers as it was to enter the family of *counting* numbers. In a few words, zero is a good thing to have … for engineering purposes where its usefulness has been proven. In the utterly prestigious world of the mathematicians, however, zero is a good thing … to be done with.

2 SCALES OF MEASUREMENT

When modern thought began to evolve in Europe in the 15th and 16th centuries, it was quite natural that astronomers, physicists, and mathematicians, began to create a jargon of their own that allowed them to communicate with one another. As time passed by and each discipline specialized in creating their own concepts and way of thinking and researching, they developed a common meta-language as it were and that was mathematics, the realm of numbers. Mathematics allowed them to understand one another not only alone from within, but also between disciplines.

Science as a whole became the queen of numbers and went much further. Scientists not only got a hold of all concepts related to astronomical, to physical, and to biological research, but they also claimed for themselves two fundamental, exclusive rights:

- The right to create instruments with which to enhance their research.

- The right to deal properly with numbers.

Towards 1900 though, there emerged a new breed of researchers that "pretended" to be able to deal with numbers too. As the body of social sciences begun to emerge and as no one could deny them the right to exist, there arose the discussion of the measurability of certain variables, human and otherwise, that one could not measure or observe by the use of a stick, a balance or a microscope; i.e. variables that were not measurable by the same means scientists were using. Social researchers had observed for instance the great differences that existed concerning the degree of intelligence amongst human beings but found themselves empty handed when it came to measuring these differences. There were no available instruments to measure them.

The new scientists felt they had no one's permission, let alone the mathematicians', first to develop a whole set of instruments that allowed them to perform their own measurements and research, and secondly to deal with numbers in their own way. In a few words, they claimed the right to found a new body of scientific knowledge alongside the existing one, that of the old science. The old scientists, not having any means to inhibit the creation of the new instruments of measurement, resented very much the use of numbers by the new breed of researchers. The realm of numbers belonged to the old science and the newcomers had better keep away from them.

As it was explained in the first chapter, the historic step that was needed to go from *counting* to *measuring* was crucial to the awareness of the existence of continuous variables, of fractions and in the end, of the number zero as it was interpreted by science. When the social sciences appeared, counting and measuring began to belong to the public domain, and the number zero as well as discrete and continuous variables became commonly known.

After a long struggle between the old and the new science there emerged an agreement about who should be allowed to use what, while dealing with numbers. The final decision centred firstly on the intensity to which variables allowed themselves to be measured, this intensity being nothing other than their ability to acquire fractional values between (any) two consecutive integers, and secondly on their ability to acquire the value of zero.

The reached agreement determined that there were 4 different scales of measurement:

- The nominal scale.

- The ordinal scale.

- The interval scale.

- The ratio scale.

Although these four scales of measurement belong to public knowledge, a short description of them shall be given here. Before entering into the matter, one should not forget that in all four cases, we are dealing with *measurements* and that all measurements, as has been stated before, have to do with numbers: measuring is equivalent to attaching numbers to the socially agreed values that a variable acquires.

First of all, the *nominal* scale: the numbers attached to the values that a variable measured at this level can acquire are actually not numbers at all. Some marketer, for instance, researches the public's preferences for green, red or yellow apples. It is a research based solely on colour. After "measuring" the preferences of thousands of people he is ready to analyse his data. As computers have less trouble dealing with numbers than with words, the researcher decides to "attach" the number 1 to "green", the number 2 to "red" and the number 3 to "yellow". As measuring is nothing other than attaching numbers to the different values that a variable can acquire, we can say that in this case a true measurement has taken place: the numbers 1, 2 and 3 have been attached to the three different values that the variable "colour of an apple" can acquire (supposing that there are only three of them), green, red and yellow.

It goes without saying that the numbers used at this level are not used as such. They are there only for convenience. Attributing numbers 1, 2 and 3 to the different colours that an apple can have has nothing to do with attributing quantitative properties to the colours they are attached to. As a matter of fact, the three numbers can be used interchangeably and indeed, each one of them could be replaced by any other number with no effect whatsoever on the final results. No wonder then that performing measurements at this level has been called *nominal*. Numbers are present only nominally, not really: they represent no numerical values at all.

Measuring at the *ordinal* level is a little more complex, but as the name itself indicates, measuring at this level has to do with *ranking*. We all learned at grammar school that numbers are to be divided into cardinal and ordinal, the former indicating quantities, the latter indicating ranking. So, *two* is a cardinal whereas *second* is an ordinal number. An example will clarify things.

A psychologist researching the development of the visual acuity of human beings decides to take measurements on 4 groups of 10 individuals each:

- 10 infants aged 1 to 3 years.

- 10 adolescents with an average of 12 years.

- 10 adults, all of the age of 49 years.

- 10 elderly with an average age of 67 years.

There are several things that are worthwhile noticing. First of all, it lies in the nature of things to rank the four groups the way we have just done, with the infants first and the elderly last. The four groups have been ranked according to age and if we decide to order them increasingly, from lowest to highest, then it becomes "logical" to attach to any group a rank higher than the previous one but lower that the next. Second, and this is extremely important, one should notice that the distances between the age groups, expressed in years, are UNequal to one another: the "distance" between groups 1 and 2 is equal to 10 years, that between groups 2 and 3 amounts to 37 years, whereas groups 3 and 4 are separated from one another by a "distance" of 18 years.

The measurement that takes place at this ordinal level consists simply of "attaching" ordinal numbers to the different values of the variable (remember: measuring is nothing other than attaching different numbers to different values of a variable) that one decides to research with one condition though: that the values that are taken are NOT equidistant from one another. In other words, at this *ordinal* level of measurement there should be some sort of chaos, although not absolute chaos.

The third measurement scale occurs at *interval* level. The difference between measuring at ordinal and at interval level is actually very simple: At the interval level of measurement the measurement points on the researched variable are taken at equidistant points from one another. Repeating, for the sake of simplicity, the last given example, the said psychologist redirects his research to these four groups:

- 10 infants all aged 5 years,
- 10 adolescents with an age of 15 years,
- 10 young people aged 25 years,
- 10 persons all of the age of 35 years.

The most important thing to know about this example is that the measurement points on the variable "age of human subjects" have been taken equidistantly from one another: they all differ by 10 years.

Finally, measurements at the *ratio* scale. This theory says: these are measurements taken on variables that admit equidistance as well as a value of zero. Trying to explain this somehow mysterious definition of the ratio measurement scale, let us go back to the classical example of measurements at *interval* level. It has been said time and again that temperature can at most be measured at interval level for these reasons: although temperature lets itself be measured equidistantly (the "distance" between 9 and 10 degrees Celsius is said to be equal to that between 10 and 11 degrees), the variable called "temperature" cannot achieve a value of zero. This point of view has been maintained even after science has found out that a zero temperature does exist (-273.15 degrees on the Celsius scale). For some reason though, science does not allow us to say that a temperature that is 2 degrees higher than -273.15 is twice as "warm" as a temperature situated only 1 degree away from the absolute zero point.

Turning back to the *ratio* scale of measurement: in order for a variable to be measured at ratio level it has to admit equidistance as well as a zero value of measurement. Length, for instance, states science, is a variable that lends itself perfectly to be measured at ratio level: one can very well think of a body of zero length, says science, but once one begins measuring length at the "right" side of zero, measuring the length of a body obeys the same laws that govern the measuring of continuous variables, as has been explained in chapter 1: a variable, in this case length, is allowed to take not only integer values but also all fractional values between any two consecutive integers.

The most vulnerable aspect of the *ratio* scale of measurement is of course its need of a zero measurement point. As it was stated in chapter 1, while "acquiring" a zero value the variable that acquires it loses its own identity. Zero light means not only no light at all, but also nothing at all: emptiness. Zero gravity means not only no gravity at all but also nothing at all: emptiness as well.

Taken as a whole, the entire discussion regarding the 4 scales of measurement was made mainly with one purpose: to distinguish old from new science. "True" science, we were told, operates only at *ratio* level; whereas the new "human" sciences were granted a free way to operate at *nominal*, *ordinal* and *interval* scale, the big difference being that ratio variables can acquire a zero-value, while ALL other variables do not, or so "old" science told us. It is however the same "old" science that, be it accidentally or not, one given day discovers that one of the most fundamental variables of our universe, temperature, reaches a point beyond which it can't decrease any further. It is an absolute zero point, one would say, determined by Nature itself. And still, says science, this nature-given absolute zero point cannot be treated as a zero point of measurement simply because it does not fit into our numerical system: a temperature of absolute 2 degrees does not represent a temperature that is twice as "warm" as a temperature of absolute 1 degree. Temperature does increase from absolute 1 to absolute 2 degrees, says science, but it does not double. The question then arises: at which point does a temperature of absolute 1 degree double? At absolute 3 degrees? At absolute 45 degrees? At absolute one million degrees? Science does not know.

The paradox about the scales of measurement is that, normally speaking, Nature does not give us free access to its own absolute zero values. This seems to occur, though, with temperature and all we have to say is: we can't handle this zero, so temperature is NOT a ratio variable, it can be measured only at interval level. On the contrary, there exist scores of variables whose absolute zero value Nature keeps hidden from us and we nevertheless pretend to be able to measure them at ratio level. Take length, which was mentioned earlier. We feel free to assign to this variable a zero-value whenever and wherever we deem it necessary. No blinking whatsoever. Besides, departing from this zero-value, we simply take a stick, lay it on the ground once, twice, three times, draw marks on the sand and proudly announce: the second mark doubles the first mark's length, the third triples it etc. etc. etc. What "on earth" makes the variable "length" that special that it allows us to manipulate and measure it any way we wish? What has "length" that "temperature" misses? Has it something to do with the physical shape of the instrument we use to perform measurements? Has it to do with the nature of the variable itself? Has it to do with the human senses through which "knowledge" of a certain variable enters into our brain? Has it to do with the nature of our numerical system? Why is a distance of two feet twice as long as a distance of one foot, but a temperature of two degrees (departing from absolute zero temperature) is not twice as warm as a temperature of 1 degree?

After all these considerations there arises a final question: suppose one day Nature decides to disclose to us the absolute zero point values of all the fundamental variables that make up our universe: zero space, zero time, zero energy ... Take then time. Set our best atomic clock running at 00:00:00. Is 00:00:02 twice as long as 00:00:01? If it is, it means that the abysmal jump from no time to time, something that had to occur between 00:00:00 and 00:00:01 had to repeat itself somewhere between 00:00:01 and 00:00:02, between 00:00:02 and 00:00:03 ... so at each new measurement point, time has to jump from nonexistence into existence; otherwise things do not add up. And if time jumps into existence each time a new measurement is done by humans, it means that time somehow had ceased to exist prior to that; otherwise, it could not jump into existence time and again. If instead of taking time, we take any other fundamental variable of our universe, be it energy, be it light, be it speed ... we are confronted with the same basic problem: once any fundamental variable of our universe comes into existence by leaving "behind" its zero measurement point, it has to come into existence time and time again whenever we humans deem it necessary or convenient to measure it. Is this what really happens in Nature? It must if we stubbornly keep alive our conviction that our numeric system is the best instrument we have at hand to interpret Nature and that a variable that really deserves this name should be able to acquire a zero measurement point.

Admitting, or rather requiring, as we do, that the fundamental variables that constitute our universe have a zero measurement point changes dramatically our conception of reality. We are then left with an intermittent universe that comes into existence time and time again at a rate that is only dictated by the speed at which the universe lends itself to be measured. And this is where the concept of measurement becomes central. Who is in charge of performing this measurement? In the present context there is but one answer to this question: the human mind. WE seem to be able to decide which variables are continuous and which are not, WE seem to be able to decide which variables must have a zero measurement point and which do not, WE are the only beings capable of making any measurement at all, so it is up to US to provide an answer as to how often either WE decide to measure the fundamental variables of Nature, or how often Nature lends itself to be measured by us.

What has just been said is what we shall call here the phenomenon of *intermittency*: while measuring the fundamental variables of Nature, admitting that each one of them possesses a zero measurement point, on our way from zero towards measurement point 1, we necessarily have to cover the abysmal gap between zero and the tiniest possible "amount" of the variable concerned. Measuring forth, we will eventually get to measurement point 2. Well, in order to arrive at 2, the "distance" covered between 1 and 2 should be equal to the "distance" between 0 and 1; for if they are unequal, we have not arrived at 2 but at some other measurement point. This means that the "distance" between 1 and 2 should also cover the abysmal gap between a non-existing and an existing variable, just as it did between 0 and 1. Well, if "leaving" 1 towards 2 means departing again from zero, this necessarily means that by arriving at 1, the variable's value automatically returned to zero, the measurement point where the variable that is being measured is nonexistent. So, while measuring, every time a measurement increases by a value equal to unity, the variable's value has to come back to zero, returning back into existence immediately thereafter. The rate of this periodicity, or *intermittency*, is determined by the variable's "amount" that we have decided to define as the variable's unity.

The absurdity of *intermittency* denies us the right to perform ratio level measurements on the fundamental variables of Nature. This compels us to conclude that we lack the basic intellectual skills that enable us to know the zero values of the fundamental variables of Nature, if indeed these points exist or ever existed. Any time we state, in the name of the human race or otherwise, to be able to "jump" over the zero value of any fundamental variable of Nature, we are behaving like gods: beings that pretend to know everything, even the "unknowable".

The general conclusion of this second chapter is not essentially different from the earlier one, taken at the end of chapter 1. If it was stated there that zero was a *fremdes*

Körper in the world of counting and measuring, here one is bound to conclude that accepting zero as a measuring point of the fundamental variables of Nature, taken in their fundamental context, leads to a bizarre universe that intermittently jumps from no existence into existence and then back into nonexistence again, this last step having the sole aim of meeting our absolute requirement that 1 plus 1 must be equal to 2, under any circumstance.

One of the ways to avoid this continuous back and forth jumping of Nature from nonexistence into existence and then back again into being could be to question the fundamental ability of our numerical system to deal properly with Nature. After all, this elementary 1, 2, 3-sequence, a humble stone-age invention, was made essentially to count goats, not the speed of light immediately after its release from absolute darkness, or the enormous amounts of energy released after two black holes have collided with one another at the other side of the universe. One goat and another goat taken together are undoubtedly two goats. Absolutely nothing stands in the way to add them up as they are totally similar. But, going back to the situation where light was two seconds old, was its second second of existence totally equal to its first second? Absolutely not. So when light was two seconds old, light was not two seconds old unless light somehow disappeared after the first second and it miraculously jumped back into existence at the beginning of the second second. This again, takes us to the intermittent universe discussed earlier. If we are prepared to accept this, then it is true that when light was two seconds old, light was indeed two seconds old. Chapters 3, 4 and 5 will hopefully throw some new light into this little black hole of the numerical system called the prime numbers.

We began this chapter with the aim of explaining the four levels of measurement that science and human sciences have agreed upon. Science has claimed for itself the exclusive right to perform measurements on variables that admit a zero measurement point. Any variable that can be measured at such a level is "theirs". Now we are ready to conclude that this is true only while we remain in the world that belongs to the engineers. If we transcend that little world and come into the realm of the fundamental variables of Nature, measurements by human beings at the ratio level are utterly impossible.

3 GENERATING THE PRIME NUMBERS

3.1 Introduction to primes

Mathematicians of all ages have been trying to find the prime numbers and have failed. Although this failure has more than one cause, the most obvious of all has been the definition itself of what a prime number is, a definition tightly bound to indivisibility. Abiding by their own rules, multiplying and dividing can be done only on continuous variables that admit a zero measurement point. It is not known whether the measuring system itself can be considered a variable or a continuous variable for that matter. Fact is that mathematicians have been studying systematically their own numerical system. Whether studying that system is tantamount to measuring it, is not known either.

So, mathematicians state that there exist a numerical sequence, or at least a numerical set that consists only of whole numbers (all prime numbers must be whole numbers) but that despite that, whatever it is that constitutes the entity of the prime numbers, that entity must be able to acquire a zero measurement point for otherwise the concept of (in)divisibility cannot be applied on it. An entity whose elements are all discrete, and therefore measurable at most at interval level cannot, even taken as a whole, acquire the properties needed to be measured at ratio level.

Being aware of this contradiction I tried to find the primes' habitat somewhere outside or rather alongside the so-called set of natural numbers. Also, after realising that looking for the primes at ratio level as the mathematicians had done for ages was nothing more than a fruitless exercise, I displaced my research in the first place to the world of the interval things. This world though imposed equidistance upon me. But equidistance, I realised, was the one single factor that posed the greatest obstacle to the mathematicians in their search for "their" prime numbers. So I abandoned that world and entered with very much hesitation indeed the world of the ordinal things.

In doing so, I must confess, I was negatively inspired by the 1001 sieves that have been invented in the past in pursuit of the primes. All sieves that I studied operate in a quite contradictory way: while discarding numbers is done at ordinal level, once a discarding wave has come to an end, a new one cannot begin without ascertaining the bare quantitative identity of the number that gives origin to the new discarding wave. To give an example: while compelled to make sure that 13 is indeed 13, the "natural" number placed between 12 and 14, NOT some n^{th} element of some sequence, numerical or otherwise, a sieve places itself outside the ordinal realm of things and enters the realm of rational measurements.

This contradictory property of all sieves led me to direct my research to a sieve that would operate entirely at ordinal level and that therefore would allow me to start looking for primes at any point alongside the sequence of the so-called natural numbers *without* having to generate all preceding primes *and* without ever having to ascertain the quantitative aspect of the numbers I researched.

Convinced as I was that (in)divisibility should be banned from the primes' essence and that being or not being a prime should be perfectly predictable in a numerical world purely arranged at ordinal level, I went ahead and searched many numerical sequences until I found one where composites could be neatly separated from primes purely on the basis of the ranking that the elements of both sub-sets held in that sequence, disregarding completely their quantitative identity.

And that is how I entered the realm of ordinal things and found the primes there, intertwined with the composites but separable from them. The sequence where both primes and composites reside, being arranged at a purely ordinal level, offers the researcher the possibility of theoretically generating all existing prime numbers, from 5 to infinity, without ever having to worry about the numerical or quantitative identity of the numbers he generates, be it composites, be it primes. Their composite or prime nature is due only to the ranking they occupy in that particular sequence where they reside side by side.

Finally, as a side-effect of arriving at the primes' ordinal hideout, there emerged the possibility of storing them in a one-dimensional array: they reside there, together with the composites, small numbers, middle-sized numbers, but also huge, huge numbers of thousands of digits, each one of them requiring not more than 1 bit to be held in a storage device. And *by all of them* I mean: from 5 to infinity.

3.2 Finding the prime numbers

Finding the prime numbers can mean many things. In this book it means: you take a rather large number, say 10,000,000. You then follow the instructions given below. The result will be that you get all prime numbers that are smaller than 10,000,000.

Instead of 10,000,000 you can choose a number that is 10 times larger, or 1000 times larger, or one million times larger i.e. 10,000,000,000,000. You will always get all primes that are smaller than any number you wish.

It is true, if you decide to use the algorithm presented in this book in order to find all prime numbers that are smaller than 10,000,000,000,000, you will have to be able to tell a computer how to do the job. But if you do, all you need is a □350-machine and give it instructions to repeat a certain operation about 1,054,091 times. I am pretty sure that within one hour or so your little machine will come up with all prime numbers that are smaller than 10,000,000,000,000.

In a nutshell the algorithm I am going to present to you does this: you generate in your computer a long sequence of zeroes. You then go ahead and check some of those zeroes (this is the exercise you should repeat about 1,054,091 times in the example given above). When you are done with the checking there are still many zeroes that have remained unchecked. Well, those are the prime numbers.

So, let us now go ahead and begin. The first thing to do is to decide about how big the largest prime number should be that you intend to generate. Suppose you decide to get all prime numbers that are smaller than about 2,000,000. Look at the word *about*. The present algorithm will allow you in due course to get exactly all primes smaller than 2,000,000. You will need some training though and the only way to get that is to train yourself by using repeatedly the present algorithm.

What I am now going to do is to explain to you how to get all prime numbers that are smaller than two different quantities: one of them will be a little bit *less* than the 2,000,000 we have just mentioned while the other one will be a little bit *more* than 2,000,000.

To achieve that, you take the next steps:

▪ Take the square root of the desired quantity. In this case: $\sqrt{2000000} = 1414.21$

▪ For reasons that will become clearer later on, you have a choice between these two squares: $1411^2 = 1,990,921$ or $1415^2 = 2,002,225$ (the 2 mentioned quantities, 1411 and 1415 should be elements of *A*, see below).

▪ Then you decide whether you want to extract all primes that are smaller than 1,990,921 or instead those that are smaller than 2,002,225. I propose that you choose 2,002,225. And by the way, I will be calling *B* this upper limit, irrespective of its size. So *B* is now equal to 2,002,225.

Once you have decided how large *B* should be, the upper limit of the primes you intend to extract, you have to generate the sequence of zeroes I was earlier talking about. But before generating it, you have to know how long that sequence should be. First though let us agree that we are going to name that sequence *Y*, and its length *n*, where

$$n = (B - 1) / 3$$

Hence you compute *n* in the example above as follows:

$$(2{,}002{,}225 - 1) / 3 = 667{,}408$$

And that is how long *Y* should be: *Y* should contain 667,408 zeroes.

If you have the skills to do it you'd better address your sequence bit-wise rather than byte-wise. This will reduce the length of the memory you need by a factor of 8 at least. If you don't know what I am talking about, simply disregard this comment. Once you generate *Y*, I propose to call its elements $y_1, y_2, y_3, \ldots y_{667,408}$.

3.3 Checking.

Once *Y* is neatly arranged in your computer you begin the checking process. What you are going to check are the elements of *Y*, the ones I have just called $y_1, y_2, y_3, \ldots y_{667,408}$

And then the checking itself. There are many ways to check something. Remember that all 667,408 elements of *Y* were set equal to zero. I suggest that whenever you check a particular element of *Y*, you simply convert it to 1.

Then there is a second rule you should keep in mind. Whenever an element of *Y* has been checked, it never gets unchecked. In other words, once an element of *Y* has been converted to 1 it never gets its original zero-value back.

Checking is done in rounds. If you look closely at Table 1 you will understand what I mean:

Clarifying the 1st round will make things clear: it means that you look for the 8th element of *Y* and convert it to 1. Then you look for the 18th element of Y and convert it to 1 as well. Then you convert the 28th element, the 38th element... all the way up until you run out of elements to convert. Arrived at that point you will have concluded your first round.

Then you begin your 2nd round: you convert the 11th element of Y to 1, then the 21st, then the 31st... you now understand what I mean.

If you have chosen a large Y, you will need far more than the 12 checking rounds you see on Table 1. (In the example given above where you were to extract all primes smaller than 2,002,225 you need 940 checking rounds to complete the job, see below). In other words, you need to know how to expand Table 1 to suit your needs. Well, this is very easily done. For that we need to analyse Table 1 a little bit closer.

Rounds						Frequency
1st	Y_8	Y_{18}	Y_{28}	Y_{38}	...	10
2nd	Y_{11}	Y_{21}	Y_{31}	Y_{41}	...	10
3rd	Y_{16}	Y_{30}	Y_{44}	Y_{58}	...	14
4th	Y_{25}	Y_{39}	Y_{53}	Y_{67}	...	14
5th	Y_{40}	Y_{62}	Y_{84}	Y_{106}	...	22
6th	Y_{47}	Y_{69}	Y_{91}	Y_{113}	...	22
7th	Y_{56}	Y_{82}	Y_{108}	Y_{134}	...	26
8th	Y_{73}	Y_{99}	Y_{125}	Y_{151}	...	26
9th	Y_{96}	Y_{130}	Y_{164}	Y_{198}	...	34
10th	Y_{107}	Y_{141}	Y_{175}	Y_{209}	...	34
11th	Y_{120}	Y_{158}	Y_{196}	Y_{234}	...	38
12th	Y_{145}	Y_{183}	Y_{221}	Y_{259}	...	38

Table 1. Checking rounds and checked Y's elements.

But before going any further I will have to clarify a couple of things.

1. In the first place, how to compute m, the number of checking rounds you need, given the upper limit B, of the primes you want to get. You get m as follows:

$m = 2(\sqrt{B} \setminus 3 - 1)$

where "\" stands for integer division.

In the example given above, if you want to get all primes < 2,002,225:

$m = 2(\sqrt{2002225} \setminus 3 - 1) = 940$

2. And secondly, the place where to find the prime numbers is a very simple numerical sequence. That sequence has the following 2 properties:

- Its first element is 5,
- It further contains all numbers that are known to you and me, except all multiples of 2 and all multiples of 3.

This is an extremely important sequence and I am going to name it sequence A. Applying the rules just mentioned, A will contain the following elements:

A = 5, 7, 11, 13, 17, 19, 23, 25, 29, 31, 35, 37, 41, 43, 47, 49...

And now that you know A (and A is really the primes hideout!), you are ready to expand Table 1. If you take a close look at the first element of the *odd* rounds on Table 1, you will see that they are closely related to A:

1st round $(5^2 - 1) / 3 = 8$
3rd round $(7^2 - 1) / 3 = 16$
5th round $(11^2 - 1) / 3 = 40$
7th round $(13^2 - 1) / 3 = 56$
9th round $(17^2 - 1) / 3 = 96$
11th round $(19^2 - 1) / 3 = 120$

And after that, look at the first element of the *even* rows:

2nd round   5 x 2 + 1 = 11
4th round   7 x 2 + 11 = 25
6th round   11 x 2 + 25 = 47
8th round   13 x 2 + 47 = 73
10th round   17 x 2 + 73 = 107
12th round   19 x 2 + 107 = 145

After seeing how easy it is to generate the very first element of each checking round, you only have to look at the last column of Table 1 to know where the frequencies come from:

1st and 2nd rounds: 5x2 = 10
3rd and 4th rounds: 7x2 = 14
5th and 6th rounds: 11x2 = 22
7th and 8th rounds: 13x2 = 26
9th and 10th rounds: 17x2 = 34
11th and 12th rounds: 19x2 = 38

As you can easily see, Table 1 can be expanded indefinitely. All you have to do is to instruct your computer to generate the first element of a round and the frequency with which the checking process has to be executed. The machine will then go ahead and execute conversion after conversion, round after round, after round... you only have to take care that the machine does not attempt to start a new round at a position beyond the last element of Y.

After all the checking is done, if you take a look at your Y, you will see that it now contains zeroes and ones, far more ones than zeroes, though. Well, the zeroes that Y now contains are the prime numbers that you were looking for when you began the whole exercise. Although, not exactly. After all, zero is not a number, let alone a prime number.

You have to look at all those zeroes in Y as being the place holders of the real prime numbers. Any two zeroes in Y, although identical, differ only in one aspect: they occupy a different place (ranking) in Y. Well, it is their ranking in Y that makes them suitable to *extract* from them the prime numbers whose place they hold in Y. If we agree to call *r* the ranking of any zero in Y, then you extract its corresponding prime number *p* from Y by stating:

p = 3*r* + 2 if *r* is odd, or
p = 3*r* + 1 if *r* is even.

Two examples will clarify this completely. You probably remember that if we want to extract all prime numbers that are smaller than 2,002,225, we have to generate Y with a length of 667,408 zeroes. Once all 940 checking rounds have been completed, if you select the zero that occupies the 1,067th position (odd) in Y and also the zero that occupies the 646,002nd position (even) in Y, this is how you convert them to their corresponding primes:

3 x 1,067 + 2 = 3,203, which is a prime number.
3 x 646,002 + 1 = 1,938,007, which is a prime number.

So, generally speaking, given a particular Y, we can say that any prime number *p* < *B* can be reduced to the following simplistic expression:

$$p = 3r + a$$

where:
- r is the ranking of a 0-element in Y after Y has been checked,
- $a = 2$ if r is odd, else $a = 1$.

It is as easy as that.

3.4 Some remarks.

I hope I kept my promise. Once Y has been checked fully you can easily convert it to A, the obscure, nameless sequence where the prime numbers reside. The whole checking of Y was done 100% at ordinal level: you were never ever, not even once, bound to look for any number's quantitative identity during the whole process. Checking was done solely on rankings.

As for the promise of finding all prime numbers, from 5 to infinity: you will of course never reach infinity but the checking process can be continued as long as you wish or as long as your resources allow you to go. Your search of prime numbers will ever come to an end, of course, due to the limitations imposed on you by your resources, not due to limitations imposed on you by the present algorithm.

Then, as promised, if you care to look at Y after it has been checked and before converting it to A you will see primes (zeroes) and composites (ones) standing side by side, together but separable.

Also, if you care to analise *A*, you will see that equidistance has disappeared. That equidistance between any 2 consecutive elements had to go became soon obvious. The final result was *A*, where an equidistance of 6 is kept between all odd and all even positions whereas the distance between even and odd positions varies from 2 to 4.

Finally, as for the promise of storing small, middle-sized and huge, huge primes in just 1 bit: all you need is a *Y* that is bit-addressable. You just go on, and on, and on, checking your *Y*. Small rankings become middle-sized rankings and later on huge, huge rankings and still, all they need to be kept in a storage device is one single bit of space. Every one of them: primes and composites.

As for the real hideout of the primes: I still don't know whether it is *A* or *Y*. Perhaps you do.

4 LARGE PRIMES

4.1 Generating primes indefinitely

While generating rather large amounts of primes, computers tend to run out of memory. The present algorithm is an exception to that in all instances but one: when you try to execute a checking round whose frequency is a number that large that your computer's memory is unable to get a hold of it. In all other instances your computer will simply go on, and on, and on generating primes as long as your processor is capable of reaching deep, deep into Y, the one-dimensional array where primes and composites reside. Nonetheless, you can and eventually you will run out of disk space where to storage Y.

So, a number that is too big to be mapped into memory and a disk space that gets depleted, these are the only physical limitations that impede the present algorithm to generate all existing prime numbers, from 5 to infinity.

Before entering into the matter though, a word of caution while choosing the upper limit B of the primes you want to extract. We all have the tendency to choose multiples of 10. We all want to see all primes that are smaller than 100,000 or than 1,000,000, or than one hundred billion. I have said it before: in due course you will be able to do it but for now you'd better restrict yourself to take the square of any element of A as the upper limit B, of your primes (remember: A contains only numbers that are greater than or equal to 5 and that are not multiples of 2 or of 3).

The **first method** to deal with this problem is this: you generate the primes step-wise, first from 5 to B, then from B to C and so on (remember: 5 is the smallest element of A). As you know by now, generating primes with the present algorithm is no other thing than checking Y. Once Y has been checked in the proper manner, the extraction of the primes is an utterly easy exercise. If you want to use this step-wise method to extract your primes, you should perform the following steps:

While generating primes from 5 to B you save in X the ranking in Y of the very last element of each checking round.

The length of X is therefore equal to m, the number of rounds you need in order to extract your primes from 5 to B. Remember though that B is the upper limit of the primes you are looking for during your first step and that therefore B should be the square of some element of A. Going back to an example I have given before, suppose you have chosen B to be 2,002,225, in other words that during your first step you want to extract all prime numbers that are smaller than 2,002,225. We have seen that the number of checking rounds you need is equal to 940, so your X should be an array that is capable of containing 940 numbers.

Once you have performed your first step, extracting your primes and saving them somewhere, you determine the new upper limit of the next series of primes you want to extract, you generate a new Y, you determine its length and you begin the checking again. You begin your checking at the position in Y that is indicated by the first element of X and, being it the first checking round, you do your checking with a frequency of 10. Once your first round is completed, you initiate your second round at the position in Y that is indicated by the second element of X and being this the second round, you do your checking with a frequency of 10 as well. A word of caution: you have to arrange things in your computer in such a way that your new Y is a prolongation of the Y you used during your first step.

At the end of each round you must not forget to save the last checked position in Y in order to facilitate things for the next step, the third one.

But coming back to your second step, once your X is depleted you have to determine the ranking of Y where your next checking round should begin. With the instructions given in the preceding chapter this should be easily done.

Once your second step is performed you begin your third step, then your fourth step... until eventually you also run out of memory or of disk space. But running out of memory depends on the way you run things on your computer. Table 2 gives an overview of the way the most important variables grow as B, the upper limit of the primes you want to extract, becomes bigger and bigger. Realising that n grows very fast gives you the chance to manage your computer's memory in an adequate manner.

For people with limited resources the best way to extract as many primes as possible is indeed to go at once from 5 to B, taking B as large as possible and then go ahead and generate all those primes in one single step. If you still have room to go further, begin again from scratch making B larger than the first time. You will eventually run out of memory or of disk space, anyway.

B	n (length of Y)	m (length of X)
25	8	0
49	16	2
121	40	4
169	84	6
289	96	8
361	120	10
529	176	12
...	...	...
2,002,225	667,408	940
...	...	...
1,000,002,000,001	333,334,000,000	666,664

Table 2. Growth of n and m as B becomes larger and larger.

The **second method** to deal with the problem is this: instead of extracting prime numbers from 5 to B (the present algorithm gives no room to extract 2 and 3 as primes, so they are not considered to be primes) you can decide to extract prime numbers that are *greater* than B instead of *smaller* than B. This method suits people with more resources and better programming skills than average.

This second method consists of *computing X* instead of *constructing* it by saving in it the ranking of the last Y-element that gets checked at the end of each checking round. And you compute X as follows:

- You determine B and B should be the square of an element of A, for instance 2,002,225 that is the square of 1,415.

- You then find the ranking r of B in Y which you find with: $r = (B-1)/3$. And I repeat: if you want to apply this method, B has to be the squared value of an element of A.

- Then you go ahead and compute X, the array that contains the places in Y where you should initiate your checking rounds. As the *frequencies* of those checking rounds (see Table 1, last column) remain unchanged no matter where you decide to initiate your checking on Y, the most important thing now to compute is the index of the first element of each checking round. You achieve this by first computing c:

$c = (r - a) \mod b$, where:

> r is the ranking of B in Y,
> a is the index of the very first element of Y to be checked when you decide to extract all primes that are greater than or equal to 5 (Table 1, second column),
> b is the frequency (Table 1, last column)

- Once you have found c, the ranking in Y where a checking round should begin is determined by:

$(B-1)/3 - c$

- By increasing a and b adequately you will be able to compute all elements of X, the array that contains the places in Y where you should initiate your checking rounds.

- Once you have computed X, you are ready to execute your whole checking process on Y as it was earlier explained.

- In order to facilitate things I have computed the first 12 checking rounds needed to find primes that are greater than 2,002,225, see Table 3. Expanding Table 3 is as easy as expanding Table 1.

- Expanding Table 3 allows you to look for primes that are greater than 2,002,225. How far you want to go is up to you but remember that B, the new upper limit has to be the square value of an element of A. For instance, letting B being equal to 152,201,569 will do as it is the square value of 12,337, a number that is divisible neither by 2, nor by 3 and that therefore is an element of A. In such a case you are looking for primes that are greater than 2,002,225 and smaller than 152,201,569.

- Once you complete your checking rounds you extract your primes from Y in exactly the same way that was explained before.

Notice that in this last example I have limited myself to find the prime numbers greater than 2 and less than 152 millions roughly speaking. It should be noticed that the second method suits better people who want to extract primes with a length of for instance 150 to 200 digits, the ones that play such an important role in cryptography (see chapter 6). If it is your intention to factor the rather big primes used by the mathematicians to seal off the Internet there is no reason for you to extract or store primes and composites in the range of 50 to 60 digits, so the present algorithm offers you the possibility of jumping directly into the range of primes, semiprimes and other composites that meet your needs.

rounds						Frequency
1st	$Y_{667,408}$	$Y_{667,418}$	$Y_{667,428}$	$Y_{667,438}$	...	10
2nd	$Y_{667,401}$	$Y_{667,411}$	$Y_{667,421}$	$Y_{667,431}$	...	10
3rd	$Y_{667,396}$	$Y_{667,410}$	$Y_{667,424}$	$Y_{667,438}$	...	14
4th	$Y_{667,405}$	$Y_{667,419}$	$Y_{667,433}$	$Y_{667,447}$	...	14
5th	$Y_{667,388}$	$Y_{667,410}$	$Y_{667,432}$	$Y_{667,454}$	...	22
6th	$Y_{667,395}$	$Y_{667,417}$	$Y_{667,439}$	$Y_{667,461}$	...	22
7th	$Y_{667,398}$	$Y_{667,424}$	$Y_{667,450}$	$Y_{667,476}$	...	26
8th	$Y_{667,389}$	$Y_{667,415}$	$Y_{667,441}$	$Y_{667,467}$	...	26
9th	$Y_{667,380}$	$Y_{667,414}$	$Y_{667,448}$	$Y_{667,482}$	...	34
10th	$Y_{667,391}$	$Y_{667,425}$	$Y_{667,459}$	$Y_{667,493}$	...	34
11th	$Y_{667,400}$	$Y_{667,438}$	$Y_{667,476}$	$Y_{667,514}$	...	38
12th	$Y_{667,387}$	$Y_{667,425}$	$Y_{667,463}$	$Y_{667,501}$	...	38

Table 3. Checking rounds to find primes > 2,002,225

4.2 Counting primes.

Oftentimes people want to know the amount of primes that are smaller than a certain number, mostly a multiple of 10. There are three ways to achieve this (beware though that the present algorithm does NOT consider 2 and 3 to be primes):

• The first one is to generate all primes that are smaller than say two billion and then count them.

• The second way is much easier: you generate Y, you check it and once you finish your checking you simply count the number of zeroes that you find in Y: that is exactly the amount of primes that you are looking for.

• There is a third way, but this one is of statistical nature. It will only approximate the result you are looking for and it is far more time consuming. It is presented here only for the sake of completeness.

Beware that in the remainder of this section I am going to talk about checking A, not Y, as I should. On the other hand Y is no other thing than a place holder of A.

Beginning with checking rounds 1 and 2, and knowing that they check 2 out of every 10 elements of A, your best guess is that these rounds will check 1 fifth of all elements between the 8th element of A and B, the upper limit of the total amount of primes that you are looking for.

Checking rounds 3 and 4 taken together check 1 out of every 7 A-elements. So, from the starting point of round 3 (and 4) you have 4 rounds running: 2 of them checking conjunctively every 5th element and the other 2 checking also conjunctively every 7th element. So, theoretically they add up checking (1/5 + 1/7) = 0.3429 % of the remaining elements of A.

But from the starting point of round 3 onwards there is a chance that rounds 3 and/or 4 will check an element that has already been checked by rounds 1 or 2, so you will have to correct for that.

Once this correction is applied it appears that the probability of an A-element being checked between the starting point of round 3 and B, the upper limit of the primes you are looking for, has to be brought back to 0.3143 %. This implies that **as far as the first 4 checking rounds are concerned**, the A-elements situated between the initial point of round 3 and B have a probability of 0.6857 % of remaining UNchecked, i.e. of being primes.

Checking rounds 5 and 6 check together 1 out of every 11 elements, you do your homework computing the probability of an element being checked that is situated after the initial point of round 5, you apply your correction, and so forth.

So, this is how you compute your pro-babilities. The starting point of every odd checking round *reduces* the probability of the remaining elements of *A* being primes whereas the subsequent correction *enhances* it. In the long run they sort of keep one another in balance, reducing very, very slowly the probability of an *A*-element being a prime without ever reducing that probability to zero.

There are many ways to try to deduce statistically the number of primes that are smaller than certain quantities. Should you work the present procedure out, then your results should be comparable to the ones below, that I took from the Internet:

Primes smaller than	Number of primes
1,000	168
10,000	1,229
100,000	9,592
1,000,000	78,498
10,000,000	664,579
100,000,000	5,761,455
1,000,000,000	50,847,534

Table 3.1 Number of primes statistically deduced.

As you see, this is a quite cumbersome procedure and I wouldn't bother using it other than to compare its results with some of the mathematicians' famous "conjectures", the ones they very much love to prove true or false. But, should you decide anyway to apply this procedure to count your primes, you should not forget to add 7 to your final results, being the first 7 primes, from 5 to 23, that were left out of the statistical procedure.

Again, a word of caution: should you decide to apply this statistical method to count your primes, you should keep in mind that you are operating on A and that A contains only about 1/3 of all so-called natural numbers. So, adapt your results accordingly whenever you leave A as your point of reference.

4.3 Some final remarks on primes.

1. In the preceding sections array A ("the mother of all primes") has been quoted repeatedly. If you are inclined to learn by intuition I suggest you take a good look at table 3.2. Notice particularly the increment of 24 that takes place column-wise over all columns.

5	7	11	13	17	19	23	25
29	31	35	37	41	43	47	49
53	55	59	61	65	67	71	73
77	79	83	85	89	91	95	97
101	103	107	109	113	115	119	121
125	127	131	133	137	139	143	145
149	151	155	157	161	163	167	169
173	175	179	181	185	187	191	193
197	199	203	205	209	211	215	217
221	223	227	229	233	235	239	241
245	247	251	253	257	259	263	265
369	271	275	277	281	283	287	289
293	295	299	301	305	307	311	313
317	319	323	325	329	331	335	337
341	343	347	349	353	355	359	361
365	367	371	373	377	379	383	385

Table 3.2 Array A arranged over 8 columns

3. In the second place it is extremely useful to look at the elements of Y that get checked as a result of the checking rounds taken two by two. So on the first row of Table 3.3 you see the places of Y that get checked during the first and second rounds, on the second row the checking of Y during the third and fourth rounds...

8	11	18	21	28	31	38	41	48	51	58	61	68	71	78	81
16	25	30	39	44	53	58	67	72	81	86	95	100	109	114	123
40	47	62	69	84	91	106	113	128	135	150	157	172	179	194	201
56	73	82	99	108	125	134	151	160	177	186	203	212	229	238	255
96	107	130	141	164	175	198	209	232	243	266	277	300	311	334	345
120	145	158	183	196	221	234	259	272	297	310	335	348	373	386	411
176	191	222	237	268	283	314	329	360	375	406	421	452	467	498	513
208	241	258	291	308	341	358	391	408	441	458	491	508	541	558	591
280	299	338	357	396	415	454	473	512	531	570	589	628	647	686	705
320	361	382	423	444	485	506	547	568	609	630	671	692	733	754	795
408	431	478	501	548	571	618	641	688	711	758	781	828	851	898	921
456	505	530	579	604	653	678	727	752	801	826	875	900	949	974	1023
560	587	642	669	724	751	806	833	888	915	970	997	1052	1079	1134	1161

Table 3.3. Checking rounds of Y taken two by two

25	35	55	65	85	95	115	125	145	155	175	185	205	215	235	245
49	77	91	119	133	161	175	203	217	245	259	287	301	329	343	371
121	143	187	209	253	275	319	341	385	407	451	473	517	539	583	605
169	221	247	299	325	377	403	455	481	533	559	611	637	689	715	767
289	323	391	425	493	527	595	629	697	731	799	833	901	935	1003	1037
361	437	475	551	589	665	703	779	817	893	931	1007	1045	1121	1159	1235
529	575	667	713	805	851	943	989	1081	1127	1219	1265	1357	1403	1495	1541
625	725	775	875	925	1025	1075	1175	1225	1325	1375	1475	1525	1625	1675	1775
841	899	1015	1073	1189	1247	1363	1421	1537	1595	1711	1769	1885	1943	2059	2117
961	1085	1147	1271	1333	1457	1519	1643	1705	1829	1891	2015	2077	2201	2263	2387
1225	1295	1435	1505	1645	1715	1855	1925	2065	2135	2275	2345	2485	2555	2695	2765
1369	1517	1591	1739	1813	1961	2035	2183	2257	2405	2479	2627	2701	2849	2923	3071
1681	1763	1927	2009	2173	2255	2419	2501	2665	2747	2911	2993	3157	3239	3403	3485

Table 3.4. Checked A-elements after checking rounds taken two by two.

And finally, on Table 3.4 you can take a look at the actual elements of *A* that get checked as a result of the preceding table.

If you care to expand these 3 tables in Excel and again, if you are inclined to learn by intuition, you will be surprised to see how much you can learn from them.

4. As for my promise of allowing you to start searching for primes at any point alongside the so-called set of natural numbers *without* having to generate all preceding primes: after reading this chapter, I hope you can agree with me that I kept my promise.

5 THE TWINS

5.1 Dealing with twins.

Before getting into the matter: all variables and concepts used in chapters 3 and 4 remain unchanged unless otherwise stated.

In order to avoid confusions though I am going to extract the twin primes from Z, a numerical sequence perfectly comparable to Y, the long zero-string we used to extract the primes from in the previous 2 chapters.

So the first thing you should do is to generate Z. In the second place you should decide where to place B, the upper limit of the twins you want to extract. And again, no matter how tempting it is to try to find all the twins that are smaller than some multiple of 10 like one million, or ten millions, or one billion, you should try to find all twins that are smaller than some number that is the square of some element of A. I suggest you take again 2,002,225, the square of 1,415, exactly like we did in the previous chapters.

Next you decide how long Z should be. Well, by now it will not surprise you that this length, *n* is found with

$$n = (B - 1) / 6$$

In this case:

$$n = (2{,}002{,}225 - 1) / 6 = 333{,}704$$

Once Z has been generated and has been given the proper length the checking can begin because yes, finding prime numbers as such and finding the twins both obey to the same principle: you generate a string of zeroes and then you check all places in Z where the twins are NOT. Once your checking is done you find the twins in those positions of Z that have remained unchecked.

As we did in the previous chapters, the checking is done in rounds. All you need to execute a particular round is to find the first element to be checked and the checking frequency.

In Table 4 you find the first 12 rounds you need whenever you start looking for the twins greater than 5 and smaller B.

As for *m*, the number of checking rounds you need in order to get all twins that are smaller than B,

$$m = 2(\sqrt{B} \setminus 3 - 1)$$

where "\" stands for *integer division.*
In the example given above, if you want to get all twins < 2,002,225:

$$m = 2(\sqrt{2002225} \setminus 3 - 1) = 940$$

At this point it is worth noticing that the first of any pair of twins has always an odd ranking in the original Y (see chapter 1), implying that the second must be even-ranked. Keeping this in mind, Table 4 was constructed where the single elements stand for the indicated pair of twins in Z. You have to keep in mind though that whereas Y contains single elements of A, Z contains **pairs** of elements, all grouped at both sides of all multiples of 6.

rounds						Frequency
1st	Z_4	Z_9	Z_{14}	Z_{19}	...	5
2nd	Z_6	Z_{11}	Z_{16}	Z_{21}	...	5
3rd	Z_8	Z_{15}	Z_{22}	Z_{29}	...	7
4th	Z_{13}	Z_{20}	Z_{27}	Z_{34}	...	7
5th	Z_{20}	Z_{31}	Z_{42}	Z_{53}	...	11
6th	Z_{24}	Z_{35}	Z_{46}	Z_{57}	...	11
7th	Z_{28}	Z_{41}	Z_{54}	Z_{67}	...	13
8th	Z_{37}	Z_{50}	Z_{63}	Z_{76}	...	13
9th	Z_{48}	Z_{65}	Z_{82}	Z_{99}	...	17
10th	Z_{54}	Z_{71}	Z_{88}	Z_{105}	...	17
11th	Z_{60}	Z_{79}	Z_{98}	Z_{117}	...	19
12th	Z_{73}	Z_{92}	Z_{111}	Z_{130}	...	19

Table 4. Checking rounds and checked Z's elements.

Table 4 can be expanded in exactly the same way as Table 2. The first element of the odd rounds and of the even rounds, as well as the frequencies, have their basis on A. Getting them should offer no difficulty by now.

After checking is completed, the twins are held in Z by proxy by those elements that have conserved their original value of 0. In order to convert all those zero-values into the twins whose place they hold, you simply look at r the ranking of that particular zero in Z and find its corresponding twins p and q as follows:

$$p = 6r - 1$$
$$q = p + 2$$

which corroborates the very well-known fact that twins *reside* at both sides of multiples of 6.

It is as simple as that.

5.2 Generating twins indefinitely.

Besides generating the twins from 5 onwards, the present algorithm offers the possibility of generating twins indefinitely, two features that no other algorithm can offer.

As it was said in the previous pages, while extracting the twins from 5 to B, the latter has to be the square of some element of A. Besides, as it was said in the previous chapters regarding the primes, it holds here also that there are two methods to extract twins indefinitely.

By means of the **first method** you generate the twins step-wise, first from 5 to B, then from B to C and so on.

If you want to use this step-wise method to extract your twins, you should perform the following steps:

• While generating twins from 5 to B you save in X the ranking in Z of the very last element of each checking round.

• The length of X is therefore equal to m, the number of rounds you need in order to extract your twins from 5 to B. If you set B equal to 2,002,225 as we did before, then the number of checking rounds you need is equal to 940, so your X should be an array that is capable of containing 940 numbers.

• Once you have performed your first step, you determine the new upper limit of the next series of twins you want to extract, you generate a new Z, you determine its length and you begin the checking again. You begin your checking at the position in Z that is indicated by the first element of X and, being it the first checking round, you do your checking with a frequency of 5. Once your first round is completed, you initiate your second round at the position in Z that is indicated by the second element of X and being this the second round, you do your checking with a frequency of 5 as well.

• At the end of each round you must not forget to save the last checked position in Z in order to facilitate things for the next step, the third one.

- But coming back to your second step, once your *X* is depleted you have to determine the ranking of *Z* where your next checking round should begin. With the instructions given on Tables 1 and 4 this should be easily done.

- Once your second step is performed you begin your third step, then your fourth step... until eventually you also run out of memory or of disk space. But running out of memory depends on the way you run things on your computer. Table 5 gives an overview of the way the most important variables grow as B, the upper limit of the twins you want to extract, becomes bigger and bigger. Realising that *n* grows very fast gives you the chance to manage your computer's memory in an adequate manner.

B	n (length of Z)	m (length of X)
25	8	0
49	16	2
121	40	4
169	84	6
289	96	8
361	120	10
529	176	12
...	...	...
2,002,225	667,408	940
...	...	...
1,000,002,000,001	333,334,000,000	666,664

Table 5. Growth of *n* and *m* as B becomes larger and larger.

For people with limited resources the best way to extract as many twins as possible is indeed to go at once from 5 to B, taking B as large as possible and then go ahead and generate all those twins in one single step. If you still have room to go further, begin again from scratch making B larger than the first time. You will eventually run out of memory or of disk space, anyway.

The **second method** to deal with the problem is this: instead of extracting twins from 5 to B you can decide to extract twins that are *greater* than B instead of *smaller* than B. This method suits people with more resources and better programming skills.

This second method consists of computing X instead of *constructing* it by saving in it the ranking of the last Z-element that gets checked at the end of each checking round. And you compute X as follows:

• You determine B and B should be the square of an element of A, for instance 2,002,225 that is the square of 1,415.

• You then find the ranking r of B in Z which you find with: $r = (B-1)/6$. And I repeat: if you want to apply this method, B has to be the squared value of an element of A.

• Then you go ahead and compute X, the array that contains the places in Z where you should initiate your checking rounds. As the *frequencies* of those checking rounds (see Table 4, last column) remain unchanged no matter where you decide to initiate your checking on Z, the most important thing now to compute is the index of the first element of each checking round. You achieve this by first computing c:

$c = (r - a) \bmod b$, where:

r is the ranking of B in Z.
a is the index of the very first element of Z to be checked when you decide to extract all twins that are greater than or equal to 5 (Table 4, second column),
b is the frequency (Table 4, last column)

• Once you have found c, the ranking in Z where a checking round should begin is determined by:

$$(B-1)/6 - c$$

• By increasing a and b adequately you will be able to compute all elements of X, the array that contains the places in Z where you should initiate your checking rounds.

• Once you have computed X, you are ready to execute your whole checking process on Z as it was earlier explained.

• In order to facilitate things I have computed the first 12 checking rounds needed to find twins that are greater than 2,002,225, see Table 6. Expanding Table 6 is as easy as expanding Table 1 (see chapter 3).

• Expanding Table 6 allows you to look for twins that are greater than 2,002,225. How far you want to go is up to you but remember that B, the new upper limit has to be the square value of an element of A. For instance, letting B being equal to 152,201,569 will do, as it is the square value of 12,337, a number that is divisible neither by 2, nor by 3 and that therefore is an element of A. In such a case you are looking for twins that are greater than 2,002,225 and smaller than 152,201,569.

Notice that in this last example I have limited myself to find the twins greater than 2 and less than 152 millions roughly speaking. It should be noticed that the second method suits better people who want to extract twins with a length of for instance 150 to 200 digits.

rounds						Frequency
1st	$Y_{333,704}$	$Y_{333,709}$	$Y_{333,714}$	$Y_{333,719}$	...	5
2nd	$Y_{333,701}$	$Y_{333,706}$	$Y_{333,711}$	$Y_{333,716}$	...	5
3rd	$Y_{333,698}$	$Y_{333,705}$	$Y_{333,712}$	$Y_{333,719}$	...	7
4th	$Y_{333,703}$	$Y_{333,710}$	$Y_{333,717}$	$Y_{333,724}$	...	7
5th	$Y_{333,694}$	$Y_{333,705}$	$Y_{333,716}$	$Y_{333,727}$	...	11
6th	$Y_{333,698}$	$Y_{333,709}$	$Y_{333,720}$	$Y_{333,731}$	...	11
7th	$Y_{333,699}$	$Y_{333,712}$	$Y_{333,725}$	$Y_{333,738}$	...	13
8th	$Y_{333,695}$	$Y_{333,708}$	$Y_{333,721}$	$Y_{333,734}$	...	13
9th	$Y_{333,690}$	$Y_{333,707}$	$Y_{333,724}$	$Y_{333,741}$	...	17
10th	$Y_{333,696}$	$Y_{333,713}$	$Y_{333,730}$	$Y_{333,747}$	...	17
11th	$Y_{333,700}$	$Y_{333,719}$	$Y_{333,738}$	$Y_{333,757}$	...	19
12th	$Y_{333,694}$	$Y_{333,713}$	$Y_{333,732}$	$Y_{333,751}$	...	19

Table 6. Checking rounds to find twins > 2,002,225

Should that be the range of the twins you are looking for, then there is no need of extracting and storing twins in numerical ranges that you don't need at all.

5.3 Some final remarks on twins.

1. Also Table 4 gets its dynamics from A.

2. Checking rounds should be brought to an end once the algorithm tries to check an element in Z that is greater than B.

3. Once an element of Z has been converted to 1 during a checking round, it never loses that status no matter how often the same element gets "hit" during subsequent rounds.

4. The dynamics of the algorithm restrict themselves to sieving all pairs of twins of which at least one is not a prime number. Sieving efficiency is fostered by the absolute absence of divisibility tests. As sieving is done on Z, an ordinally arranged sequence, the sieving process can be prolonged at wish, without slowing down, no matter how far one decides to go into $\mathbb{N}$. The process is particularly fast if it is decided to represent Z in memory bit-wise, not byte-wise.

5. Once again, you should keep in mind that in this chapter twins are referred to as if they were one single unity. To take an example, the 5th element of Z is the pair [23,25] which by the way is not a twin.

6. The possibility to start looking for primes at any point alongside the so-called set of natural numbers will show how twins do behave as researchers dig deep into that set without first having to generate all preceding twins.

6 CRYPTOGRAPHY

6.1 Introduction.

There are a few brands of cryptography that love to play this hide and seek game on the Internet: "I will give you a semiprime, you go ahead and find the two primes whose product the said semiprime is. If you can". The mathematicians concerned, having chosen rather large primes for their game, bet that you could never ever find those two primes and went ahead and secured the Internet on that assumption. You could prove them wrong, who knows. You just go ahead and try it.

There are two methods to try to solve the mathematicians' problem. The first method is the classical one: you go ahead and try to find one of the prime factors whose product the semiprime is. Once one of them has been found, finding the second prime factor is trivial.

The second method does not look for the prime factors but goes directly after the mathematicians' semiprime. Or rather, you go directly to the *place* in Y where the semiprime should be, a place that originally belonged, and actually still belongs exclusively to the concerned semiprime. And you find that *place* in one trillionth of a second. But you have fooled the mathematicians: instead of finding there their semiprime you have cared to save there in advance one of the prime factors the mathematicians asked you to find. Finding the second prime factor is trivial as you undoubtedly have already deduced.

6.2 First factorising method.

Before entering fully into the matter, let us agree on some names: let us call S the mathematicians' semiprime, s the square root of S, p_1 the smallest of both primes and p_2 the largest one.

In order to solve the mathematicians' game you have to access a portion, or sub-set of Y that has already been properly checked. But then a double question arises: why do you need a properly checked Y and which portion, or sub-set thereof is indispensable for you in solving the problem we have at hand.

To begin with the first question: if you are in the business of solving the mathematicians' game we are talking about (also known as RSA or public key cryptography), one of the ways to achieve your goal is to be in possession of a data base of some sort where to look for p_1 and/or p_2.

Well, in these pages we have been dealing a lot with A, a numerical sequence that contains not only all prime numbers but also all products of any pair of prime numbers. In other words, A contains not only the source of the problem, being the mathematicians' semiprime but also the solution to that problem, being the two prime factors you are looking for, p_1 and p_2.

This is A. But what about Y? Well, at the beginning of chapter 3 we have seen how easy it is to generate Y from scratch on the one side and how easy it is to convert Y into A on the other side. As a matter of fact Y and A are conceptually indistinguishable from one another. They are both ordinal sequences where all that counts is the ranking of their elements.

I would imagine that if you deal with primes, you had better generate them once and for all and keep them somewhere instead of having to generate them time and again every single time you need them. And this is where Y comes into focus, for a properly checked Y is by far the best place to keep your primes ready to go, the more so if your Y is bit-addressable.

So, with Y in the background you can access any subset thereof, or all of it for that matter, any time the need arises. You then convert it to A the way it was explained in chapter 1 and you are ready to go.

And then the second question: which portion, or subset of Y, and consequently, of A do you need in order to find p_1 and/or p_2.

The answer to this question is extremely difficult. While committed to solve the mathematicians' problem, all you get from them is a semiprime while it is your task to find one of its prime factors. Theoretically you do a good job whenever you select that portion of Y where, after Y has been converted into A, you can easily find the prime factor you are looking for, be it p_1, be it p_2.

The first step you should take is to compute s, the square root of the semiprime. And in order to make things easier you should take an s such that,

$$s = \sqrt{S} \setminus 1$$

and you apply the integer division by 1 in order to get rid of the decimals.

Once you have got s, you know that $p_1 < s$ and that $p_2 > s$. You also know that in order to solve the problem at hand you need to find only one of both primes. In the third place, due to the fact that numbers are what they are, you also know that the distance between s and p_1 is nearly always numerically shorter than the distance between s and p_2 and this is why you always should try to find p_1, instead of p_2.

You then should go ahead and find p_1 there where it can be found: in A. But there are several ways you can scan A. An example will clarify things. Suppose you get from the mathematicians the semiprime 555,991 and you are asked to factorise it. As it was earlier stated, the first thing to do is to find s. In this case you compute:

$$s = \sqrt{555991} \setminus 1 = 745$$

Arrived at this point you will have to make sure that s is an element of A. If it is not, you should replace it by the nearest A-element that is greater than s. In this case it is not necessary as 745 is not divisible by 2 or by 3 and is therefore an A-element.

By now you know that $p_1 < 745$. But 745 is the 248th element of A. Should you scan A from low to high, beginning with 5, A's first element? Or, on the contrary, should you scan A from high to low, beginning with 745, A's 248th element? Or should you scan A beginning at some other point?

It is a matter of public record that mathematicians for the game at hand tend to choose a semiprime whose two prime factors are not too far from one another on the numerical scale. In other words, in the example given you are better off if you begin scanning A from high to low, beginning at 745.

For the sake of saving some space, I am now going to reproduce a sub-set of A horizontally instead of vertically, as I am used to do in Excel:

A= 5, 7, 11, 13, ... 599, 601, 605, 607, 611, 613, 617, 619, ... 739, 743, 745, ...

I am now supposing you are going to scan A from high to low, beginning at 745, so first you compute:

555991 / 743 = 748.31

555991 / 739 = 752.36

555991 / 737 = 754.40 ...

Arrived at this point, seeing the way the decimal part of the quotient behaves and being you looking for an integer quotient with no decimals, there exist all sorts of shortcuts you can use in order to speed up your search. For this, I refer to the Internet.

But continuing with the explanation, I am going to suppose your scanning of A goes ahead, dividing 555991 by every single element of it until a particular trial division yields an integer quotient with no decimals. This happens when you arrive at:

555991 / 613 = 907

At which point you have solved the mathematicians' problem. It has taken you 44 trial divisions to get there. The primes you were looking for are 613 and 907.

The **second way** to solve the problem presupposes that you scan only those elements of A that are primes. After all, you know in advance that p_1, and p_2 are both primes so why bother dividing the semiprime by a divisor of which you know that it is not a prime. Since we already know now that p_1 = 613, I am going to list all concerned primes:

613, 617, 619, 631, 641, 643, 647, 653, 659, 661, 673, 677, 683, 691, 701, 709, 719, 727, 733, 739, 743.

I am now supposing that you have saved these primes in Y and that you scan them, again from high to low, beginning with 743 (745 is not a prime) So, this way of doing things reduces the number of trial divisions you need from 44 to 21, an exercise that is quite worthwhile executing. This supposes though that you have extracted and saved in Y (or in A, whichever you prefer) the sub-set of primes you need, in this case a substantial part of all primes that are smaller than 745.

The **third way** to solve the problem is very similar to the second one, where you scan the concerned primes from high to low until you eventually arrive at p_1, in this case 613.

This third way implies that instead of scanning the concerned primes in descending order, you first randomise the natural ranking order of a number of them, prior to your scanning. Notice though that I have said *a number of them*. Nobody will be able though to tell you how many primes you should randomise. In the example given, *if* you were to randomise the first 21 primes smaller than 745, you will reduce your search substantially *in the long run* for 613 will be the last prime to be scanned only in 1 out of every 21 trials. So in 20 out of every 21 trials you will have speeded up your search in varying degrees. There is no way though to know that you optimise things by randomising exactly the first sub-set of 21 primes that are smaller than 745. In most instances you will undoubtedly choose another number and this will deteriorate your chances.

As you undoubtedly will conclude, the chances of solving the mathematicians' problem discussed thus far are heavily correlated with the amount of primes you can extract and save after you have checked Y in a proper manner. Resources and an inventive mind are the key realities that will eventually lead you to solve the problem. Y (of A if you wish) is a powerful sequence that will lead you to extract and save as many primes as your resources allow *and* to extract and save them in any desired range of the so-called natural numbers. It is therefore my conviction that, given identical resources to those that the mathematicians have at their disposal, if you have a bright mind you will be able to beat them at their own game. And I do hope you will. Should you after all remain sceptical then I strongly recommend that you study the second factorising method explained below.

6.3 Second factorising method.

6.3.1 Numerical background.

The second method to solve the mathematicians' problem is heavily related to what I shall call the *t*-value of a particular element of Y.

Referring to chapter 3 you will remember that the first step to find the primes consisted of simply creating a long sequence of zeroes, that we called Y. You then checked Y time and again bringing into play what we then called the *checking rounds*. Once the checking process had come to an end you were left over with a fully checked Y that now contained zeroes and ones, by far more ones than zeroes. You may also remember that the zeroes in Y were no other thing than the place holders of the prime numbers you were after whereas the ones occupied the place of all composites. At that time it was indeed my suggestion that you should check Y in the way that has just been described.

For the second factorising method that I am now about to describe, I suggest that you check Y in a slightly different manner.

In the first place, the whole theory about checking rounds, starting points, frequencies, ... as these have been described in chapters 3 and 4 should be maintained. The only thing that changes is the way you execute the actual checking on Y. In chapter 3 it was suggested that you convert into a 1 every zero that was "hit" during the checking process and that once an element of Y had been converted to 1, it never ever should regain its original value of zero.

The change I am now going to suggest is this: during the checking process, if the element of Y that gets "hit" during a particular round has still the value of zero, just go ahead and convert it to 1. However, if that element has already a value of 1, convert it to anything else provided it is neither a zero nor a 1. In other words, if you check Y in this way, you reserve the value of zero for those elements of Y that never get checked (these are the primes, as you know) and the value of 1 for those elements of Y that get checked only once (and these are the semiprimes).

Bur before continuing with the explanation of the second method as such, I hope you will allow me to make a suggestion that could bear fruits should you decide to deepen your knowledge of the prime numbers using the methodology exposed in these pages. Well, the suggestion is this: if you check Y the way I have been explaining thus far in this paragraph, once your checking of Y is done you are left over with a Y that contains place holders for the primes (zeroes), place holders for the semiprimes, those composites that are the product of two primes and of two primes only (ones) and place holders for the composites that are the product of three or more primes (and it was up to you to decide which number or symbol to use as place holder). What I now suggest is that you accumulate in Y itself the number of times a particular element gets "hit" during the whole checking process. This means that each time you "hit" an element of Y you increase by 1 whatever value it is that already "resides" in that element, so at the end of the day your Y will tell you the number of prime factors into which *any* composite can be factorised.

But going back to the second factorising method, you could perhaps agree with me that if you check Y in the way that has been suggested, your Y tells you the number of **times** that each one of its elements gets "hit" during the whole checking process. This is why I decided to call this the *t*-value of each element of Y.

Well, this second factorising method relies fully on this *t*-value. Remember that the mathematicians' game consists of challenging you to find the two hidden primes whose product, their semiprime, they make public on the Internet. Well, a key to find those two primes consists of a proper arrangement of all composites that have a t-value of 1 for these are the semiprimes. On the face of it, it is an almost impossible task. However, looking at it more closely you will see that it can be done.

In the first place and reducing our scope to the range of *Y*-elements (or *A*-elements if you wish) beneath 1,000, just take a look at the list of primes that fall within that range (remember: the numbers 2 and 3 are **not** considered to be primes in these pages).

```
  5   7  11  13  17  19  23  29  31  37  41  43  47
 53  59  61  67  71  73  79  83  89  97 101 103 107 109
113 127 131 137 139 149 151 157 163 167 173 179     181
191 193 197 199 211 223 227 229 233 239 241 251
257 263 269 271 277 281 283 293 307 311 313 317 331
337 347 349 353 359 367 373 379 383 389 397 401 409
419 421 431 433 439 443 449 457 461 463 467 479 487
491 499 503 509 521 523 541 547 557 563 569 571 577
587 593 599 601 607 613 617 619 631 641 643 647 653
659 661 673 677 683 691 701 709 719 727 733 739 743
751 757 761 769 773 787 797 809 811 821 823 827 829
839 853 857 859 863 877 881 883 887 907 911 919 929
937 941 947 953 967 971 977 983 991 997
```

Table 7. List of all 166 primes < 1,000.

If you care to count, there are 166 of them. You will remember that while explaining the *first* factorising method the most efficient way to find p_1 and p_2 relied on you being in possession of a data base that contained all primes in the range where you expected to find find p_1 and p_2.

Well, if you take the same range of A-elements beneath 1,000 and you look at the number of elements with a *t*-value of 1 (semiprimes) that it contains, keeping in mind that the same numerical range contains 166 primes, you can have a first idea of the efficiency of this second method as compared to the first one.

25	35	49	55	65	77	85	91	95	115
119	121	133	143	145	155	161	169	185	187
203	205	209	215	217	221	235	247	253	259
265	287	289	295	299	301	305	319	323	329
335	341	355	361	365	371	377	391	395	403
407	413	415	427	437	445	451	469	473	481
485	493	497	505	511	515	517	527	529	533
535	545	551	553	559	565	581	583	589	611
623	629	635	649	655	667	671	679	685	689
695	697	703	707	713	721	731	737	745	749
755	763	767	779	781	785	791	793	799	803
815	817	835	841	851	865	869	871	889	893
895	899	901	905	913	917	923	943	949	955
959	961	965	973	979	985	989	995		

Table 8. List of all 138 semiprimes < 1.000.

So, if you look at Tables 7 and 8 you see that the number of semiprimes that are smaller than 1,000 is less than the number of primes that are smaller than 1,000. Besides if you look at Table 8 a little closer you will see that it contains more than a few *A*-elements that are multiples of 5 as well as a couple of squared quantities.

When trying to solve the mathematicians' problem, you must take into account that they will never ever publish on the Internet a semiprime that has a prime smaller than 1000 as one of its factors.

Anyway, we could strip Table 8 of all semiprimes that are multiples of 5 as well as of all squared quantities. Such an operation yields the following results:

~~25~~	~~35~~	~~49~~	~~55~~	~~65~~	77	~~85~~	91	~~95~~	~~115~~
119	~~121~~	133	143	~~145~~	~~155~~	161	~~169~~	~~185~~	187
203	~~205~~	209	~~215~~	217	221	~~235~~	247	253	259
~~265~~	287	~~289~~	~~295~~	299	301	~~305~~	319	323	329
~~335~~	341	~~355~~	~~361~~	~~365~~	371	377	391	~~395~~	403
407	413	~~415~~	427	437	~~445~~	451	469	473	481
~~485~~	493	497	~~505~~	511	~~515~~	517	527	~~529~~	533
~~535~~	~~545~~	551	553	559	~~565~~	581	583	589	611
623	629	~~635~~	649	~~655~~	667	671	679	~~685~~	689
~~695~~	697	703	707	713	721	731	737	~~745~~	749
~~755~~	763	767	779	781	~~785~~	791	793	799	803
~~815~~	817	~~835~~	~~841~~	851	~~865~~	869	871	889	893
~~895~~	899	901	~~905~~	913	917	923	943	949	~~955~~
959	~~961~~	~~965~~	973	979	~~985~~	989	~~995~~		

Table 9. List of all 86 useful semiprimes < 1,000.

As you can see, you are left over with only 86 out of 138 semiprimes. All I intend to do with this little exercise though is to familiarize you a little bit closer with the semiprimes since you are going to need them badly.

6.3.2 Theoretical background.

Before getting fully into the matter, a clarification is needed: the relationship between the so-called natural numbers, a fully checked Y and A.

The "natural numbers" are that curious sequence 1, 2, 3, ... we all learn at grammar school. They may not be "natural" but they are certainly *numbers*. Science and engineering tell us we can manipulate them at will and that is exactly what we do: we multiply them, we subtract them from one another, we square them ... For you it is important to know that the semiprime the mathematicians publish on the Internet is an element of their numerical sequence, a sequence that I am going to call N.

As for Y, this is the array that originally consists of an endless row of zeroes and that is the object of the whole checking process as this has been extensively explained in the last chapters. Both before and after completion of the checking process Y retains its name.

A, like Y, is an ordinal sequence. And I remind you once again that the only thing that matters to the elements of an ordinal sequence is their *ranking*. In fact, an ordinal sequence can contain anything and should 2 such sequences have an identical number of elements, then they are completely exchangeable with one another: the third elephant in an ordinal sequence of elephants can be swapped and/or replaced with the third element of a sequence of oranges. That is why A and Y are conceptually identical to one another irrespective of whether Y has or has not yet been checked.

However for the sake of clarity I have tried to keep A and Y apart from one another keeping in A numbers that look very much like elements of N and in Y most of the time zeroes and ones. As for how to convert Y into A, I refer to the previous chapters where such conversion has been repeatedly executed.

The method though that I am about to explain to you requires your mental ability to navigate freely from and towards anyone of these 3 sequences, A, N and Y.

Beware that the numbers you see on Tables 7 to 9 are elements of A, not of Y, even though for the mathematicians they are all elements of N. Besides, you certainly will remember that the checking rules given in chapter 3 were meant to be applied on Y, not on A. The mathematicians ask you to find p_1 and p_2, both elements of A, not of Y. The mathematicians though think of p_1 and p_2 as being elements of N.

The methodology applied in these pages concerning the generation of prime numbers and twins differs profoundly from the way the mathematicians look at primes and twins. For them, twins, primes and semiprimes are all elements of N whereas in these pages those 3 entities are and remain mere rankings, elements of an ordinal array. So, this is what you should keep in mind: we are going to use ordinal entities to try to solve a problem put to us by people that can think only in terms of quantities.

This second factoring method relies heavily on a data base, a fact common to both methods. Whether the construction of such a data base is feasible will be discussed later on. For the moment let us suppose it is feasible and go ahead.

In theory it is easier to find an object x that is an element of a small set of entities than to find an element y that is element of a large set. My theory then is that while trying to solve the mathematicians' problem, taking into account the size of the semiprime published by them on the Internet, one is better off chasing the semiprime than chasing the two prime factors. And the reason is obvious: in the numerical range where the relevant primes and semiprimes must be found there are far less semiprimes than primes so you will find the semiprime earlier than the primes whose product the semiprime is.

But then you ask yourself: why to bother looking for a semiprime that you already know? The reason is this: the semiprime we get from the mathematicians is a quantitative entity whereas we work in an ordinal field. And ordinal entities are not numbers. They are rankings, as we have seen. Things will become clear as we go on.

6.3.3 Applying the method

Let us for a moment go back to a fully checked Y, in which the "ones" are the semiprimes' place holders. Let us pretend we are going to check Y again using checking rounds, starting points, checking frequencies, ... everything as it was explained earlier. With one exception though: this time we are going to check only the "ones", leaving unchanged all other elements of Y.

By "checking only the *ones*" I mean this: you just go ahead and execute the checking rounds on Y pretending Y was never checked before. Every time you "hit" a position on Y you will have to take a look at what you "hit". If it is not a "one", leave that position alone, don't do anything. But if it is a "one" apply the checking described in the next paragraph.

The checking of the "ones" consists in wiping them out of their place and replacing them by a number that is equal to one half of the current frequency. At this point it is worthwhile remembering that the checking rounds, taken two by two, have exactly the same frequency: the first **and** the second rounds have a frequency of 10, the third **and** the fourth rounds have a frequency of 14, ... so if a "one" is "hit" during the execution of the third round, it should be replaced by one half of 14, being 7.

And arrived at this point we have reached the heart of the second factorising method. A sequence Y that has been checked in this way is actually the data base you need in order to solve the mathematicians' problem. The way it works will be explained with a couple of examples.

In the first place let us suppose that we got from the mathematicians the semiprime 799, one of the numbers you see on Table 9. For the mathematicians, 799 is just a cardinal number. For us in this place 799 is an element of A. The only thing relevant to 799 is its ranking in that sequence. But being A and Y perfectly interchangeable, we compute the semiprime's ranking in Y as follows:

$$799 \setminus 3 = 266$$

Arrived at this point, all you have to do is to go ahead and take a look at the 266th element of Y. The "one" that was there has disappeared though. It was replaced by 17 during the last checking you did on that sequence. Apparently this was done during the execution of a checking round that had a frequency of 34: it must have happened while you were executing the 9th or the 10th checking rounds both of which have a frequency of 34. Well, 17 is one of the prime factors the mathematicians asked you to look for. The other prime is or course:

$$799 / 17 = 47$$

So, this is the essence of this second factorising method: you grab the mathematicians' semiprime from the Internet, you calculate its position in Y, you go and take a look at that position in Y and there it is: one of both prime factors is there, waiting for you to be grabbed. It is as simple as that. A quicker answer to the mathematicians' challenge is unimaginable.

I am now going back to an example given earlier, where you were supposed to have got from the mathematicians the semiprime 555,991, an element of *A* as you know. The first thing you should do is to find the semiprime's ranking in Y:

$$555991 \setminus 3 = 185330$$

So, the mathematicians' semiprime occupies the 185330th position in Y. You go ahead and take a look at that position where you see 613 waiting for you. It must have been placed there while you were executing the 407th/408th checking rounds, both of which have a frequency of 1226. The second prime factor is of course:

$$555991 / 613 = 907$$

A last example: should you decide to factorise a semiprime using the so-called *Lenstra elliptic-curve factorisation* method, you are almost bound to factorise

455,839, a number that has become sort of sacred. Don't ask me why. Mr. Lenstra's method looks really impressive on the Internet. But if you factorise 455,839 in exactly the same way you just did with 555,991 using the simple, nameless method presented here, you begin with:

$$455839 \setminus 3 = 151946$$

implying that 455839 is Y's 151946th element. You look at that position in Y and find there 599, one of the prime factors. The other one is:

$$455839 / 599 = 761$$

And again there they are: after looking at Mr. Lenstra's sacred number's position in Y, you have found 599 and 761, the two primes that multiplied by one another yield the mathematicians' semiprime. It doesn't look as impressive as Mr Lenstra's method, one has to concede, but it works, and it works extremely fast. And it has taken you maybe one trillionth of a second.

6.3.4 Data base feasibility.

Both the first and the second factorising methods explained in these pages presuppose the existence of a data base that you can consult while trying to solve the mathematicians' problem.

It is very well known that the data base attached to the first method is not feasible, should it be intended to factorise even moderately large semiprimes due to the sheer superabundance of primes in our counting system.

The feasibility of the data base that goes with the

second factorising method seems to be quite another matter. This does not imply that anyone, notwithstanding his/her brilliant mind but with no more resources than a laptop computer at his/her disposal can just go ahead and construct a data base apt to meet the mathematicians' challenge. Not being a computer expert, I am not in a position to judge who is and who is not in a position to constitute a real challenge to the forces that operate in the market place sealing off the Internet. Nonetheless I dare think that both the computer and the human resources present even in mediocre universities, of the type that are abundant in my own country, should be sufficient to challenge the forces that dominate the market place.

Before entering fully into the matter I should explain in some detail the way to navigate between *A* and *Y* on the one side and *N* on the other. As numbers become larger and larger in *N*, one way to express figures shortly is to represent them as powers of 2 or of 10, something that cannot be done with elements of *A* or of *Y*. Also semiprimes are quoted in *N*, not in *A* or in *Y*. On the other side, checking is quoted in *Y* and so do frequencies, starting points and checking rounds.

The reader should keep in mind that *A* (and *Y*) is 3 times shorter as compared to *N*, due to the fact that all multiples of 2 and of 3 were banned from *A* (and consequently form *Y*). So, whenever we state that a checking round has a frequency of 94 for instance, this means that the checking frequency on *Y* advances 94 positions on that sequence after a particular place has been "hit". Transposed to *N* though, a frequency of 94 on *Y* is tantamount to a step of 282 positions on *N*. Keep this well in mind as we are going to deal with rather large numbers that are going to be expressed in *N* as powers of 2 or of 10.

Computing their checking frequencies and starting points is done mostly in *Y*. So, please, don't get confused.

The first thing to keep in mind while determining whether the data base needed for this second factorising method is feasible is to remember how the semiprimes' places get marked on *Y*. Keep this in mind though: I am now talking about the **first time** you check *Y*, when you were advised to reserve the "zeroes" for the primes, the "ones" for the semiprimes and any other symbol of your choice for higher-order composites. Well, a semiprime is a position in *Y* that gets "hit" only once during the whole checking process, a position that held a "zero" before getting "hit". If a position is not checked at all it houses a prime number and if it gets checked 2 or more times, it is the place holder of a composite that is the product of 3 or more prime factors.

The feasibility of a data base in which to house the semiprimes is strongly related to the occurrence pace of semiprimes as *N* expands. If the probability of such an occurrence grows as *N* expands, the sheer numbers will impede the construction of such a data base. If on the contrary that probability decreases as *N* gets bigger and bigger, then the possibility of creating such a data base could become real.

One first symptom is given by the relationship that exists between the number of primes and of semiprimes that are smaller than 1,000,000. Well, consulting Table 3.1 we see that the number of primes is equal to 78,498 whereas the number of semiprimes is equal to 13,695. This last number is computed as follows: you have seen on Table 7 that the number of primes < 1000 (in *N*) is equal to 166. As a semiprime is the product of any one of those primes multiplied by another prime, their total number is equal to:

$$(166^2 - 166)/2 = 13{,}695$$

So, even in the lowest region of N where the percentage of primes is high we can see that the primes outnumber the semiprimes by more than 5 times. We shall see that as N expands the primes outnumber the semiprimes, not just by 5 times as we have just seen, but by billions and billions of times.

In Y, the probability of a zero-position getting "hit" decreases as the number of checking rounds increases due to two reasons: in the first place because the more rounds get executed the less zeroes are left over in Y to be passively "hit" and in the second place because the checking frequency increases as the checking process goes on and this diminishes the probability of a checking round actively "hitting" a zero-position in Y. So, as N expands the lesser zeroes are left over to be "hit" and the lesser instruments are left over to actively "hit" positions in Y.

Being it so that by the time N gets as big as 10^{12} the percentage of primes is about 3,76% of all elements of N, a checking round that then begins to get executed has a probability of only 3,76% (measured on Y though this probability is equal to 11,28%, 3 times larger) of hitting a zero, that then is a candidate to be a semiprime. But it is only a candidate for it runs the chance of getting "hit" during the execution of a posterior checking round, losing its status of semiprime.

But anyway, disregarding for the moment this last possibility, and knowing that a checking round whose execution gets initiated not far from 10^{12} has a frequency of 2,000,000 implies that the first time this round "hits" Y is 6,000,000 away from 10^{12} (a "distance" of 2,000,000 on Y, where the frequencies are measured, implies a distance 3 times as large on N as it was earlier explained). On the other hand, knowing that in this range the primes are not more than 11,28% of all the elements of Y, implies that the checking round being executed has the probability of "hitting" a zero only once every 9 times it "hits". Disregarding the probability of this semiprime losing its status due to another "hit" during the execution of a posterior round one can conclude that at this stage a numerical bracket of 54,000,000 on N (9 x 6,000,000) contains 2,030,400 primes but only 1 semiprime.

Applying the same reasoning to 10^{24} leads to the conclusion that a checking round whose execution begins not far from 10^{24} has a frequency of $6(10^{12})$. Lowering to a conservative 2,75% the percentage of primes present in N (8,25% in Y) in that numerical range, one has to conclude that going as far as 10^{24} it takes a numerical bracket of $72(10^{12})$ in N to contain 1,980,000,000,000 prime numbers but only 1 semiprime.

The conclusion is that by the time one reaches 10^{200} the semiprimes have become a rarity, really hard to find. If one considers that semiprimes in the range of 2^{512} with their 250-260 digits have been used, and are still used to seal off the Internet, one wonders what happens if the mathematicians go over to the 2^{1024}-range with its 300-plus digits. They are certainly better off sealing off the Internet from primes-hunters. But do they also protect the Internet effectively from semiprimes-hunters? Who knows.

I hope to have demonstrated not only that the presence of semiprimes in N decrease as N expands, but also that their presence decreases dramatically. It is not up to people like me to utter a judgment about the overall feasibility of a data base as it has been described in the previous pages. You don't have to be a computer expert though to be allowed to assert that computer people fix anything feasible. If they do that in this case, they will prove me right when I decided to give this book the title I gave it: *RSA Cracked*.

6.3.5 Final thoughts on the second method.

The second factorising method is heavily focused on the search for semiprimes instead of primes. Theoretically, these two ways to try to solve the mathematicians' problem have advantages and disadvantages. On the face of it, searching for the primes is seemingly easier to do: there are two of them and the factors to be found tend to be half as long as the semiprimes are.

Searching for the semiprime though has the great advantage that you do know what to look for. If your data base is all right, factoring the semiprime occurs in a blink of an eye or, as I have repeatedly said in these pages, *in one trillionth of a second*. Also, you can study in advance the sort and size of the semiprimes that the mathematicians tend to publish on the Internet and this gives you the opportunity to tune up the numerical bracket in N (or Y) that is most suitable to give them an instantaneous answer.

Presupposing that the number of semiprimes present in N is not an obstacle for the construction of a data base, there are still some other problems that wait to be solved. The first of them is tied to this question: at which point in Y should you start your checking. Most certainly not at Y's 8th position as Table 1 suggests. Fortunately, in chapter 4,

Table 3 you are given the opportunity to start your checking at any point in Y. The effort needed is great and the resources you need in order to execute this are considerable. Giving an example, if it is your intention to factorise semiprimes as big as 10^{100}, you could start your checking at say, 10^{99}.

Another problem is posed by the size of your data base. As long as Y remains bit-addressable, disk-space management remains theoretically feasible. In the proposed data base though the "ones" have been wiped out and have been replaced by rather huge numbers. No being a computer expert I dare not propose any solution to this problem even though I can imagine more than one.

Finally, as for the ability to find huge numbers in your data base: theoretically you will have no problem in finding them as long as you keep them in the world of ordinal things. It is their place in Y that should be addressable, not their quantity in N. In the ordinal world you simply search for positions, not for quantities. And whether the number concerned consists of 2 or of 300 digits, is completely irrelevant.

6.4 Some final remarks on cryptography.

The dramatic decrease of semiprimes in the world of numbers as N expands was to me totally unexpected. Apparently, one of the worlds where ordinal things reside (A or Y, you tell me), makes a sharp distinction between positions that do not accommodate to a certain predictability, being the primes, and positions that do accommodate to that rhythmical periodicity, being the composites.

Prior to exploring the behaviour of semiprimes in the numerical system as N expands, I intuitively thought of some sort of Gauss-distribution in which, given a fixed upper limit, the amount of primes shouldn't be distinguishable from the amount of semiprimes to be found there. After all, primes are "composed" of zero factors and semiprimes are composed of the least possible number of factors, namely 2. They seem to have contiguous values, so to speak.

After witnessing the dramatic decrease of the presence of semiprimes in Y as N expands, I am inclined to think that it is the composites that have some sort of normal distribution and that the primes are positions in the ordinal world that are completely left out of that distribution.

If it is true that such a normal distribution exists, and that the semiprimes have the smallest value on that distribution "on the left side", one keeps wondering about the greatest possible value that such a distribution can have, about its mean, its standard deviation, ...

Before expressing my guess about this matter I should explain something: going back to chapter 3, you will see that the checking *frequencies* change every 2 checking rounds and that the initial point of all odd checking rounds practically coincides with the successive squares of the elements of A. Well, my guess is that in the long run, if you take all positions on A between any two consecutive odd rounds, the composites are normally distributed with the semiprimes having the least occurrence "on the left side of the curve". As for the other parameters of such a distribution, not having at my disposal neither the resources nor the knowledge to make measurements "on the spot", I dare not say anything.

In order for such a measurement to be somehow representative it has to be done on a sufficiently large sample of positions in *A*. I intuitively think that 10,000 positions in *A* (30,000 in *N*) should give a fair indication of what happens in the long run. Well 14,999 and 15,001 are two consecutive *A*-elements whose square values differ by 29,998 (in *N*), so one should be able to count the number of primes that exist between 224,970,001 and 225,030,001(in *N*), and to look, in the same numerical bracket, at the distribution of composites "weighted" according to the number of prime factors they are the product of. I wonder.

7 REFLECTIONS

From what has been written in the first two chapters, there are many conclusions to be drown and many questions to be asked. By far the most important of all the questions is: how reliable is our counting system when it comes to using it to represent and to measure Nature?

Science has no doubt achieved remarkable results and will continue to do so as long as we continue to be the dominant species on Earth and its surroundings. We are great engineers. The problem of science though is multiple:

- It has used uncritically a counting system invented by our ancestors thousands of years ago that was aimed at measuring discrete variables. Science, however, has made use of the same system to measure continuous variables.

- Once continuous variables had come into existence, science introduced the number zero, placing it at a distance from 1 that was supposed to be equal to the distance between any two other consecutive numbers.

- The extraordinary engineering achievements obtained by using such a continuous counting system with the number zero attached to it led science to believe that Nature's fundamental variables were ordered in accordance to our Stone-Age counting system as it had been enhanced by science, and to believe that we therefore were entitled to measure Nature with it.

The greatest fallacies of this way of thinking are:

- Nature does not admit a zero measurement point in its fundamental variables. While measuring, numbers are supposed to be marking points on any scale used to measure variables, be they discrete or be they continuous. Well, at "zero point" any variable has long ceased to exist, except for engineering purposes where any point, mainly in the time/space continuum, can arbitrarily be marked with the number zero.

- While measuring the fundamental variables of Nature, *equidistance* between the number zero and the number 1 on the one hand, and between any two other consecutive numbers of the counting system on the other, is false; for it leads to Nature's *intermittency*.

- Zero is as far from the smallest number we can imagine as "infinite" from the largest possible number we can conceive. So zero is as unconceivable to the human mind as "infinite" is. They are both unreachable to our brain.

- Accepting zero as a measurement point in any fundamental variable of Nature is incompatible with its *measurability*, measurability being nothing other than the repeatability of the chosen measurement unit on the ratio measurement scale. Departing from the zero point, one never ever reaches point 1 no matter how tiny the measurement unit is, for reaching such a point leads necessarily to the possibility of reaching measurement point 2, and this reduces any variable to intermittency, as we have seen.

Finally the big question arises: what then is science's number zero? In these pages it has been suggested that such a point is a sort of convergence of all variables, both discrete and continuous. At zero energy, there is not only an absolute absence of any energy, but also of anything else: it is an absolute vacuum. The same holds for zero time, zero space or zero goats for that matter. It is not even the departing point of any of those variables, it is the absolute absence of anything at all.

That zero is an absolute vacuum becomes clear considering this: suppose for a moment that zero mass contains indeed no mass at all, without being an absolute vacuum. If the latter is the case, then zero mass should contain at least the tiniest possible remains of any other fundamental variable of Nature, be it time, be it energy ... In this case, the fundamental step from zero mass to mass does NOT take place as mass does NOT originate from zero mass but from time, from energy ... from the tiniest possible remains of any variable that was present at zero mass, at least if zero mass was not equal to absolute vacuum.

Earlier in this book the question has been put: how reliable is our counting system when it comes to using it to represent and to measure Nature? Well, we must say that although it has a great reliability for engineering purposes, it is completely unreliable when it comes to using it to measure the fundamental variables of Nature. And the reason is: zero, as a number, is hidden from us. Measuring time, space, energy ... has necessarily to begin *after* such variables have come into existence, once they have left their zero measurement point behind. Only from then on are we allowed to begin measuring them with our Stone Age counting system enhanced with infinitesimal steps between any two consecutive marking points. In a few words, deprived of their zero value, Nature's continuous variables are measurable by the human mind only at *interval* level. This should surprise no one considering for instance what happens to temperature: as stated earlier, we have apparently been able to find this variable's lowest possible value at -273.15 degrees Celsius. With this Nature-given jewel in hand, why are we still unable to establish a measurement unity that enables us to make measurements at ratio level on temperature? And then, extrapolating to time: suppose one day Nature reveals to us zero time. Our clocks then start running. How can we be sure that the time elapsed between 00:00:00 and 00:00:02 is twice as long as the interval that runs between 00:00:00 and 00:00:01? And then, supposing a 2-second interval is twice as long as a 1-second interval: where does this certainty come from? Is it because time reveals itself to us better than temperature does, or is it because our brain is better equipped to understand time than it is to understand temperature? Why are we not able to find in temperature a unity that makes it fit to be measured at ratio level whereas with time *any* chosen time-length will do?

Science has been claiming for the past centuries to be able to measure Nature from the zero point on, including the number zero itself. What is then this zero measurement point science has claimed to have found?

It should be noticed that science's number zero has the following properties:

- It is supposed to exist.

- Any variable's zero measurement point is supposed to be different from any other variable's zero measurement point.

- There is continuity between the zero measurement point and the least possible, measurable value of the variable being measured.

- The gap between the variable's non-existing point (its zero measurement point) and the least possible measurable value of the variable concerned occurs only once, namely right after the variable has left its zero measurement point. Nonetheless the variable is measurable, and any measurement unity that is chosen does not lead to intermittency.

So, we have to deal with a zero point:

- that does exist.

- that originates all fundamental variables of Nature.

- that diversifies all those fundamental variables of Nature at the zero measurement point, i.e., even before they become measurable to the human mind.

- that gives origin once and for all to all fundamental variables of Nature making them measurable, and impeding intermittency.

In trying to identify the essence of science's number zero, there is only one solution: the Number Zero is God, the origin of everything, the only being able to diversify Nature before its fundamental variables come into existence, the only being capable of creating time, not from zero time but from absolute nothingness, capable of creating space, not from zero space but from absolute nothingness, capable of creating light, not from zero light but from absolute nothingness ... a being that is capable of discerning between the nothingness that gave birth to time, the nothingness that gave birth to space, the nothingness that gave birth to light ... But most important of all, this is not the God preached by any religion or by all of them for that matter: this is the God whose existence is necessarily postulated by the most fundamental laws of Nature ... as these have been conceived by science itself.

Converting science's Number Zero to God, the Origin of Nature, changes dramatically the essence of the fundamental, continuous variables of Nature. Deprived of their zero measurement point, time, space, energy, light ... become floating variables as it were, with no beginning and no end. This should surprise no one: measuring, as has been said time and again, is nothing other than attaching numbers to variables. Well, numbers are all terms of our numerical system, a system that certainly has no end, but a system too that has no beginning even though we have thought for centuries to have found this beginning at point zero.

Deprived of their "roots" in the number zero, what do these variables still represent? Knowing that under such circumstances, time, space, energy, light ... can be measured only *after* they have come into existence, i.e., after they have left their zero-status, at which point of their own continuum do these variables lend themselves to allow the human mind to start measuring them? But irrespective of the position of that primordial measurement point, one should never forget that we are then measuring light in *already* existing light, space in *already* existing space, time in *already* existing time ... and besides, looking at the words *after* and *already*: why is it that we human beings are equipped to conceive of a variable's coming into existence only within the fundamental variable of time? Deprived of its zero origin, we can conceive space's coming into existence only in time, not in space itself, not in energy, not in mass. All this is due to our ability to conceive of *any* change as occurring *only* in time. A change of the fundamental variable of time that takes place in the fundamental variable of energy is as unconceivable to us as is a change of the fundamental variable of light that takes place in the fundamental variable of light itself, or in mass, or in gravity. All these changes, so we are told by our brains, can occur only in time.

So, looking at any fundamental step from zero energy to energy, from zero light to light, from zero gravity to gravity ... our brain dictates that such a step can occur only in time. But generally speaking, *any* change that takes place in *any* fundamental variable can be conceived by human beings only as taking place in time. If a change occurs for instance in gravity, the only way for this change to occur "outside time" is to conceive time as a non-existing variable at the "moment" such a change takes place. But for us human beings, a change that takes place outside time is inconceivable. In such a case, we can only think of the situation "before" and the situation "after". On the other hand, if such a change takes place while time is already "running", the only way for such an event to take place "outside time" is that the situation prior to the change and the situation after the change occurred simultaneously, something our brain can't conceive; for in such a case, a particular situation should have to have existed and not existed at the same time.

What is so fundamentally wrong with our brain that we are bound to make time-dependent the very origin of all fundamental variables of Nature? Or is time the origin, not only of time itself but also of all fundamental variables of Nature? Or, on the contrary, is time not one of Nature's fundamental variables at all?

Be that as it may, we are equipped to know the fundamental variables of Nature as being distinct from one another: time is not energy, mass is not light, light is not time. We can of course *speculate* that they are all just one variable, that there are no fundamental variables at all, ... but that is not the way our brain was made to know Nature. So unless Nature has been playing games with us, the fundamental variables of Nature *are* distinct from one another. Why Nature has forbidden us to *measure* those fundamental variables outside time, we will probably never know.

After all these considerations, what happens to the scales of measurement? If we stick to the four measurement levels explained in chapter 2, then we are bound to conclude that while trying to "measure" Nature's fundamental variables, we can't go further than the *interval* level as those variables, deprived as they are of their zero value point, do not admit measurement at ratio level.

At interval level though, equidistance is still needed and this requires a constant unity. The problem with unities is that although we are free to choose them, we have no other choice than to fix their size based on the knowledge we have at present of the fundamental variables of Nature, ignoring what might have happened to them right after they had left their zero-status if indeed they ever had such a status. To science, this is the more ominous as it pretends to be capable of conceiving those variables *at* their zero measurement point. Be that as it may, should the fundamental variables of Nature ever have come into existence, it could well be that those variables, undergoing the abysmal changes that then were taking place, were unfit to be measured by the unities we conceive now. How long did time take to become time? Did gravity come from absolute heaviness or indeed, from zero weight? Did light originate from absolute brilliantness or on the contrary, from absolute darkness? Did all these dramatic changes take place in time or was time not yet "there"? What does a second or a minute measure while time is still becoming time?

All in all then, discarding Nature's measurability at ratio level, the best we can do, as it has just been said, is to measure the fundamental variables of Nature at *interval* level with all sorts of unbelievable consequences, the most important of all being: the impossibility of carrying out ratio operations, not only within, but also between them. If time cannot be multiplied by time, if light speed cannot be squared, if mass is not divisible by energy ... What happens for instance to the mother of all equations:

$$E = mc^2 \ ?$$

One cannot discard the possibility that one day an extremely clever engineer will show that taking a discrete amount of mass and multiplying it by the square of the speed of light produces a discrete amount of energy. Whether the resulting energy will express itself in temperature, in light brightness, in sound, in gravity ... it really does not matter. When that occurs, one should bear in mind that the taken matter did not originate at matter's absolute nil measurement point and that the resulting energy did not give origin to energy in Nature as both matter and energy had long before, in time, come into existence.

So, coming back for a moment to Mr. Einstein's equation, one has to conclude that if there was mass to be taken, not as a constant but as a fundamental variable of Nature, multiplying that mass by the squared speed of light would lead not only to an incommensurable amount of Energy but also to contradictorily exclusive amounts of Energy, should Nature decide to repeatedly "re-create" Energy from Mass. How great will those amounts of Energy be, we simply are not equipped to know. Perhaps Nature is. After all: what does 12 degrees Celsius multiplied by 7 degrees Celsius yield? Any answer will do as no answer can be proved false AND no answer will do as any answer can be proven false. We live in an uncertain universe no matter how much we would prefer to dwell in a certain, predictable one.

Having arrived at this stage, is it not strange that, as chapter 3 suggests, prime numbers leave their mysterious, hidden places once they are placed in a numerical sequence that has been arranged, not in an *interval*, (let alone in a *ratio*) but in an *ordinal* way? Rather than being numbers that are divisible only by themselves and unity, are prime numbers not those that do not fit in a particular sequence that has been arranged in an *ordinal* way? Is *indivisibility* the most essential characteristic of prime numbers, or is it *ranking*? Are primes "foreign" numbers in a *ratio* world or in an *ordinal* one? Why are prime numbers, all of them, blindly localisable in a numerical sequence that does not even meet the simple rules that mathematicians impose to their arithmetic progressions?

And dedicating my last words in this book to the prime numbers, I shall state that seeing the way they behave at the surface of the ordinal world, I am inclined to think that there is something fundamentally wrong with the way we count things.

Oftentimes the primes are described as the building blocks of our numerical system. There is no reason for me to consider myself to be an authority on any field of human knowledge, but supposing for a moment that primes are not quantities but mere *positions* in an ordinal sequence as I have been repeatedly stating in these pages, how on earth can they be considered to be the building blocks of a numerical sequence whose elements are used to multiply and divide, to add up and to subtract to say the least? How can we multiply 35 by 35 if the "building blocks" of this quantity, 7 and 5 are no other thing than the first and the second element of an ordinal sequence? What does multiplying "first" by "second" yield? They yield 35 only if they, besides being ordinal, are also quantitative. Or, are they not ordinal and only quantitative? But if they are only quantitative, how come there is no way to find the primes, all of them, in that quantitative world? How come there is no way to isolate semiprimes, all of them, from primes and from higher-order composites in that quantitative world whereas even an ignorant is capable of doing that, effortlessly besides, provided he enters the word of things ordinal?

Looking at Y (= 0,0,0,0,0,0,0,0,0,0, ...) is like looking at a box full of mysteries. All its elements are identical. Their only difference is their ranking. And it is their rankings the entities that give you access to the quantitative world of the mathematicians. Any constant you use and any operation you execute on them will deliver you a different numerical sequence. Most significant in these pages has been the transformation of *Y* in *A*. Key to get *A* from *Y* has been the number 3. Multiply each ranking in *Y* by 3 and add 2 or 1 to the result depending on whether the ranking is odd or even, yields *A*.

But then, once you have got *A*, this sequence dictates how should you go back to *Y* and transform it in order to enable you to get **from Y itself** at will any amount of prime numbers, any amount of semiprimes, any amount of higher-order composites. Why is such fine-tuned, endless operation impossible to execute in the quantitative world? It is as if the ordinal world (*Y*) encounters in *A* the quantitative world of the mathematicians, dispossesses it of 0 and 1, its most dangerous numbers and rids it of all its multiples of 2 and of 3. Only then seems *A* to be fully operational, ready to bring you back to the ordinal world, a world where primes and composites reside. Side by side but separable.

Having said this, I dare not state that primes and or composites belong to the world of the mathematicians. Perhaps you do.

THE AUTHOR

German Navarro has authored several novels in Dutch, his own language. Besides, he has written two novels in Spanish and a book on spirituality, published in Italian. The present work is his first publication in English. German Navarro is Dutch, has lived and worked in several countries, one of them being Italy. Rome in particular, the former capital of the world, has been his habitat for the past decade or thereabouts.

In *RSA cracked* the author invites the reader to reflect on some aspects of the fundamental variables that constitute our universe: time, mass, energy, light ... Are they what they seem to be? But most importantly, are they what science tells us they are? If the author is right, there is something fundamentally wrong both with the way science, including scientists like Mr. Einstein himself, claim to be able to measure our universe AND with the capability of our brain to understand the very nature of Nature.

To the reader interested in *prime numbers*, the author offers in this book an original solution to that problem. Not only individual primes but also twins, dispossessed of their